AF790840

Optical Properties of Graphene in Magnetic and Electric Fields

Optical Properties of Graphene in Magnetic and Electric Fields

Chiun-Yan Lin
Thi-Nga Do
Yao-Kung Huang
Ming-Fa Lin
Department of Physics, National Cheng Kung University, Tainan 701, Taiwan

IOP Publishing, Bristol, UK

© IOP Publishing Ltd 2017

All rights reserved. No part of this publication may be reproduced, stored in a retrieval system or transmitted in any form or by any means, electronic, mechanical, photocopying, recording or otherwise, without the prior permission of the publisher, or as expressly permitted by law or under terms agreed with the appropriate rights organization. Multiple copying is permitted in accordance with the terms of licences issued by the Copyright Licensing Agency, the Copyright Clearance Centre and other reproduction rights organisations.

Permission to make use of IOP Publishing content other than as set out above may be sought at permissions@iop.org.

Chiun-Yan Lin, Thi-Nga Do, Yao-Kung Huang, Ming-Fa Lin have asserted their right to be identified as the authors of this work in accordance with sections 77 and 78 of the Copyright, Designs and Patents Act 1988.

ISBN 978-0-7503-1566-1 (ebook)
ISBN 978-0-7503-1564-7 (print)
ISBN 978-0-7503-1565-4 (mobi)

DOI 10.1088/978-0-7503-1566-1

Version: 20171201

IOP Expanding Physics
ISSN 2053-2563 (online)
ISSN 2054-7315 (print)

British Library Cataloguing-in-Publication Data: A catalogue record for this book is available from the British Library.

Published by IOP Publishing, wholly owned by The Institute of Physics, London

IOP Publishing, Temple Circus, Temple Way, Bristol, BS1 6HG, UK

US Office: IOP Publishing, Inc., 190 North Independence Mall West, Suite 601, Philadelphia, PA 19106, USA

Contents

Preface

This book intends to illustrate the development of the generalized tight-binding model in graphene systems. The contents cover the AA, AB, ABC, AAB and sliding configurations and discuss many factors affecting their magneto-electronic properties. The absorption spectra of multilayer graphene systems are investigated in detail under the various stacking configurations, and their competition and cooperation with the applied electric and magnetic field are also studied. The main spectral characteristics focus on the prominent structures, number, intensity and frequency of the absorption peaks. In chapter 2, we introduce the Peierls tight-binding model under electric and magnetic fields. The optical response, which is used to describe the spectral properties, is calculated based on Fermi's golden rule. We start with monolayer graphene in chapter 3, and subsequently discuss AA-stacked trilayer graphene. Symmetry-preserved multi-Dirac-cone structures are responsible for the optical absorption spectra. Intra-Dirac-cone excitations comprise the whole frequency region of the spectra. Chapters 4 and 5 are devoted to AB- and ABC-stacked trilayer graphene. The excitations of massless and massive Dirac quasi-particles dominate the former system, while the vertical transitions between any two inter- or intragroup Landau levels (LLs) are permitted for the latter. In particular, the phenomenon of LL anticrossings in the energy spectrum is discussed and explained using a clear physical picture, especially for the energy width of the sombrero-shaped subbands. This leads to the corresponding non-monotonic dependence of the spectral frequency and intensity on the magnetic field. In chapters 6 and 7, we discuss the optical response under the effects of broken inversion symmetry in AAB-stacked trilayer and sliding bilayer graphene. The low-symmetry stacking configurations trigger more significant LL anticrossings and cause distinct features for the K- and K′-splitting LLs.

IOP Publishing

Optical Properties of Graphene in Magnetic and Electric Fields

Chiun-Yan Lin, Thi-Nga Do, Yao-Kung Huang and Ming-Fa Lin

Chapter 1

Introduction

Graphene is a 2D material made up of hexagonal carbon lattices [1, 2]. Since mono- and few-layer graphene sheets were first fabricated in 2004 [1, 2], low-dimensional graphene-related systems have been of great interest in experimental and theoretical studies. The stacking order of graphene sheets includes the essential sequences of AA [3, 4], AB [5–9], ABC [5, 8–11] stacking. While the AA stacking configuration has only been made artificially from intercalated graphite compounds, the AB and ABC configurations are common orders in natural graphite, with their respective estimated volume fractions of 80% and 14% [15, 16]. The remaining ~6% consists of haphazardly stacked graphene sheets in a turbostratic configuration [17–19]. Few-layer graphene with the desired specific stacking configuration can be exfoliated from highly orientated pyrolytic graphite [1, 2] as well as chemically and electro-chemically reduced from graphene oxide [20–27]. Nevertheless, the chemical vapor deposition method has the advantage of producing large-scale, high-quality graphene sheets. Recently, a large area of graphene with high mobility and highly symmetric configurations, e.g. AA, AB and ABC, was found in chemical vapour deposition (CVD)-grown samples [28–40]. The improved quality is adequate for research experiments and industrial applications [41–49]. In addition, AAB stacking and intermediate bilayer configurations, with relative layer-layer shifts or twists, have also been found and are becoming interesting subjects [59–67]. Synthesized graphene is generally comprised of different kinds of domain due to the tiny difference between the total energies of the different configurations. Infrared spectroscopy, transmission electron microscopy (TEM) [3–5, 34], scanning tunneling microscopy (STM) and spectroscopy (STS) [6–9, 11, 35–37], angle-resolved photo-emission spectroscopy (ARPES) [68–74], and Raman spectroscopy [10, 75, 76], can be used to visualize the stacking domains with nanometric resolution for cases of tri-, tetra- and pentalayer graphene with typical or intermediate configurations.

Monolayer graphene belongs to a gapless system which possesses extremely high carrier mobility of up to 15 000 $cm^2\ V^{-1}\ s^{-1}$ [1]. P R Wallace discovered that the low-energy spectrum around the corners K and K′ in the Brillouin zone is described

doi:10.1088/978-0-7503-1566-1ch1

© IOP Publishing Ltd 2017

by isotropically conical dispersions in the framework of the nearest-neighbor tight-binding model [77]. The conical band structure and the vertex points at K and K′ are known as the Dirac cone and Dirac points, respectively. The linearly dispersed quasi-particles in graphene behave as relativistic Dirac fermions with an effective speed of light $\simeq c/300$ (c is the speed of light). Few-layer graphene with specific stacking configurations, e.g. AA [78–80], AB [81–93], ABC [93–99] and AAB [64], are predicted to display unique electronic energy dispersions. This new material holds great promise for the development of next-generation electronic and optoelectronic nanodevices [41–47, 50–56] because the electronic and optical properties can be flexibly tuned by the application of external fields [64–74] as well as changes to the geometric structures [75–114] and dopants [115–122]. The stacking configurations have been experimentally and theoretically verified to have great significance for the magneto-electronic and optical properties [64–67, 108, 124–189], shedding some light on other intriguing structure-specific physical properties, e.g. the Coulomb excitations and the quantum Hall transport properties [190–221]. The essential properties under previous investigations are rather different from those in other layered materials, conventional metals, semiconductors and topological insulators [222–235]; they are dominated by the interplay between the external fields and the specific geometric symmetries. In this paper, we thoroughly review the electronic and optical properties of multilayer graphene to explore its response under external fields with respect to the specific AA, AB, ABC, AAB and sliding configurations.

In AA-stacked graphene, all the graphene sheets are identically stacked along the $\hat{z}$ direction [126, 153]. The low-energy band structure consists of several monolayer-like massless Dirac cones [78–80, 153], which are exactly located at the two valleys K and K′, while the energies of the Dirac points shift away from the Fermi energy due to the vertically projected geometry. The absorption spectrum is, in general, a superposition of monolayer-like spectra, because only the vertical excitations of intra-Dirac-cones are permitted, regardless of the external electric and magnetic fields [153, 183]. However, the stacking effect gives rise to an optical gap for the even-layer configurations, which is a forbidden optical transition zone resulting from free carriers in the Dirac cones. The middle-frequency spectrum shows symmetric peaks as a result of the saddle-point excitation channels around the M point. Such peaks are also observed for other stacked graphene, as well as being a critical characteristic of sp^2-bonding carbon systems. In a uniform perpendicular magnetic field $B = B_0\hat{z}$, the Landau level (LL) spectrum behaves in the same way as several groups of monolayer-like spectra [126, 153, 183]. Responsible for each Dirac cone, the quantized LLs of each group are characterized by a set of single quantum modes, and their energies can be described by a square root dependence on B_0. Therefore, the magneto-absorption spectrum consists of several monolayer-like spectra with delta-function-like absorption peaks following the optical selection rule of $\Delta n = \pm 1$ [153, 183]. A forbidden transition region and a discontinuous threshold frequency would be revealed for all the bilayer-like intragroup LL transitions, as the Fermi level ($E_F = 0$) deviates from the Dirac point. Moreover, the preservation of the optical and electronic properties of the massless Dirac fermions indicate the unchanged spatial symmetry of the chemical environment under the external fields.

AB-stacked graphene is a fascinating material which has attracted intense attention due to the nature of its massless and/or massive Dirac fermions, depending on whether the numbers of their stacking layers are even or odd [82–93]. Of special interest is the electronic structure, which can be essentially characterized by the superposition of monolayer-like and bilayer-like structures. Optical excitation channels accessible to the monolayer-like and bilayer-like subbands construct the optical spectrum [81, 100, 154–157]. Under an electric field, bilayer graphene shows a sharp peak due to new transition channels within the energy gap (up to 100 meV) [102, 103, 107, 109], and for cases in which there are more than three layers, displays enhanced semimetal characteristics attributed to the significant overlap of the low-energy subbands [101, 109, 110, 114]. In particular, as a result of the nearly isotropic energy dispersions around the K point, quasi-1D asymmetric peaks are revealed for the vertical excitations between two constant energy loops. Electric-field-induced characteristic peaks are also observed in other stacking systems. On the other hand, a magnetic field gives rise to interesting Landau quantization phenomena in the electronic and optical properties of odd-layer symmetry-broken graphene and even-layer inversion-symmetry graphene [126, 128–131, 141, 142]. Massless and massive Dirac quasi-particles reveal single- and twin-peak structures, respectively, in the monolayer-like and bilayer-like spectra [65, 158]. The corresponding ratio of the peak intensities is estimated to be about 2:1. In the presence of an electric field, the magneto-excitation channels around K and K′ are separated from each other due to LL splitting, so that the channel numbers become double those under zero electric field [126]. In addition, very few LLs with a special quantum mode relationship avoid crossing each other during the variation of the magnetic or electric field; these are characterized by the main quantum mode and the side modes under the hybridization of the LLs [126, 131, 149]. Extra optical selection rules are induced for such LLs with multi-quantum-modes in addition to the ordinary $\Delta n = \pm 1$. In certain regions of the absorption spectra, the effects of geometry breaking and the anticrossings of intergroup LLs would trigger some special peaks without the simple relationships between the intensity, frequency and field strength.

Few-layer ABC-stacked graphene is a semimetal [93–99]. The electronic structure is characterized by one pair of partially flat subbands near $E_F = 0$, and pairs of sombrero-shaped subbands near the energy of the vertical atomic interactions between the nearest-neighboring layers [93–99]. Contributed by the surface-localized states [97–99], the partially flat subbands give rise to a prominent peak in the density of states (DOS) and play an important role in the low-frequency optical spectrum [110, 112]. An electric field can separate the two partially flat bands and trigger a new kind of optical channel responsible for a gap transition [110, 112]. Under a magnetic field, the intergroup and intragroup excitation channels among well-behaved LLs comprise the main part of the absorption spectrum [146, 174–176], in which the absorption peaks show a twin-peak structure because the LL energy spectrum is asymmetric about $E_F = 0$ [175, 176]. Moreover, it is interesting to study the quantization effects of any two constant energy loops in the sombrero-shaped structure, where the quantized energies of the LLs are proportional or inversely proportional to the magnetic field strength [126, 146, 147]. This leads to many anticrossings of intragroup LLs in the ABC stacking configuration. By varying the electric field strength, one can suppress or enhance the degree of LL anticrossings and change the components of the main mode and side modes, that is, one can induce and modulate optical channels for progressive LLs during the variation of the field strength. Furthermore, the absorption spectrum is no longer dominated by a twin-peak structure but rather by a double-peak structure under a sufficiently large electric field—a transition which is attributed to the spatial symmetry breaking of the LL subenvelope functions.

AAB-stacked graphene exhibits an extraordinary band structure compared to the AA, AB and ABC systems, mainly due to the low symmetry of its stacking geometry [64]. In the case of the trilayer system, AAB-stacked graphene is a narrow-gap semiconductor which presents a pair of subbands with strong oscillatory energy dispersions near $E_F = 0$. With increasing energy, pairs of sombrero-shaped and parabolic bands appear in substantial amounts. In the low-frequency region, rich optical excitation channels cause many absorption peaks, some of which show quasi-1D asymmetric divergence due to the vertical excitations between two constant energy loops. The LLs coming from the two valleys K and K′ are obviously separate in the mirror symmetry broken system [64]. The results present certain important differences for the state degeneracy, spacing, quantum number and spatial distribution of the wavefunctions, as compared with the highly symmetric configurations. Furthermore, the low-symmetry geometry of the AAB configuration implies the ease of triggering complex patterns of LL anticrossings and the richness of the magneto-absorption spectra. At low energies, the oscillatory subbands can induce triple-degenerate Landau states under the condition of three constant energy contours. In the B_0-dependent LL spectrum, there are many prominent multicrossings and anticrossings, i.e. the existence of both well-behaved and perturbed subenvelope functions, and sometimes abnormal ones resulting from severe LL hybridizations. Accordingly, intergroup and intragroup LL anticrossings are both shown for various regions, and they even coexist in a narrow region. The optical characteristics resulting from either the intragroup or intergroup inter-LL excitations are remarkably different for the K- and K′-valley splitting states. The main reason for this is the phase and amplitude of the LL wavefunctions.

An intermediate bilayer configuration is formed from AA to AB, and then to AA′, as the two graphene sheets shift relative to each other in the armchair direction [66, 148]. The two Dirac cones transform into parabolic bands from AA to AB configurations, and for a further layer-layer shift, they tilt towards each other at a small angle, restoring the AA′ configuration. It should be noted that the two Dirac points are located at different wave vectors and energies once the stacking configuration deviates from the AA configuration. Furthermore, an eye-shaped stateless region is formed near $E_F = 0$ as a result of the strong hybridization of Dirac cones. The dramatic transformation between the Dirac cone structure and the parabolic band structure leads to drastic changes to the optical response under external electric and magnetic fields. The layer-layer shift has a significant impact on the LLs in sliding bilayer systems. In addition to the well-behaved and perturbed LLs stemming from the massless and massive Dirac fermions in the highly symmetric AA and AB configurations, some undefined LLs appear to be promising in the case of a significant mixture of LLs, as the geometry deviates from the typical configurations. However, there is an absence of any regular optical selection rule for such undefined LLs, because their wavefunctions display extremely irregular oscillations without a dominating quantum mode. On the other hand, when the AA′ configuration is built, the LLs become well-behaved again, while additional optical channels and selection rules are accessible for intragroup LL excitations within the tilted Dirac cones. In the case of different bilayer configurations, it would be worthwhile to investigate the considerate shift-induced absorption peaks, which appear in a wide range of anticrossing patterns in the field-dependent energy spectrum.

The electronic properties are studied by use of STM, STS [6–9, 11] and ARPES [10, 68–76]. The band structures based on the tight-binding model agree with the first principles calculations [64, 66, 77, 78, 80–84, 89, 90, 93, 94, 96, 148]. The isolated Dirac cones are clearly seen in the monolayer graphene [69, 71, 73, 74], and the stacking effects for the typical configurations, AA, AB and ABC, are also identified by the main characteristics of the band structures [78, 80, 82, 126]. The former two show multiple Dirac cone dispersions and the superposition of monolayer-like and bilayer-like dispersions respectively [8], while the latter is not analogous to the AA and AB configurations [68, 70]. On the other hand, the deformation of Dirac cones for the AAB- and sliding bilayer configurations [64, 66, 148] can be verified by using ARPES. The aforementioned electronic characteristics have also been confirmed in absorption [164, 166–168, 173, 174, 177], transmission [180, 181], refection [136], Raman scattering and Rayleigh scattering spectroscopy [75, 76, 138, 139, 171, 172, 236–243]. The evidence of Dirac cones has been observed for monolayer graphene in terms of the zero threshold frequency and the linear dependence of the intensity on the frequency. The low-frequency spectral features have also been verified for the AB- and ABC-stacked samples on SiO_2/Si substrates. Furthermore, the results provide spectroscopic evidence of a tunable gap in biased AB-stacked bilayer and ABC-stacked trilayer graphene through the observation of new gap-induced absorption peaks, with a blue or red shift at different gate voltages [109, 110]. The gap size is compatible with the theoretical calculations based on the tight-binding model and *ab initio* theory [100–106]. Transport and STS measurements also give the modified electronic properties under the electric field, including the enhanced semimetal properties of AB-stacked trilayer graphene [108, 112–114]. The electric-field-induced excitations rely on the geometric structure; that is to say, verifications of the spectral structure would be worthwhile in further research to identify the interlayer atomic interactions and geometric structures.

The quantized LL energies directly correspond to the prominent DOS peaks measured in STS, where the LLs near $E_F = 0$ have been identified for monolayer, AB-stacked bilayer and trilayer as well as ABC-stacked trilayer graphene [244–254]. Quantum Hall transport measurements also observe the sequence of the LLs [203–220]. Furthermore, low-energy optical transitions with specified selection rules are verified for the intergroup LLs near $E_F = 0$ in the infrared transmission, absorption and magneto-Raman spectra [75, 76, 136, 138, 139, 164, 166–168, 171–174,

177, 180, 181, 236–243]. The spectral features of the equivalent monolayer-like and bilayer-like Landau quantization are, respectively, identified by the square root and linear dependence of the energies on the magnetic field strength in the infrared transmission of AB-stacked trilayer graphene and in the magneto-Raman spectroscopy of AB-stacked graphene up to five layers [171]. For ABC-stacked trilayer graphene, while quantum Hall effect (QHE) transport measurements show the low-lying LLs [217–219], experimental evidence on the inter-LL excitations—distinct from the massless and massive Dirac fermion excitations—is needed for further verification. The crossings and anticrossings of LLs enrich the essential magneto-electronic and optical properties [64–66, 126, 131, 135, 146, 147, 175]. The former has been confirmed by the shift of plateaus in QHE measurements due to the increase in the LL degeneracy [214–220]. Regarding the latter, one can use optical spectroscopy to identify the non-monotonic changes in the structures, frequencies, intensities and numbers of the absorption peaks [175]. This phenomenon is expected to be more significant in low-symmetry AAB-stacked and sliding bilayer graphene [65, 66]. In sufficiently low magnetic fields, the spectral features reflect the band structures, such as the symmetry, energy dispersions and atomic interactions, as well as the broken mirror symmetry under the electric field. These experimental results can be used to identify the stacking configurations and the tight-binding parameters.

In this work, the generalized tight-binding model is developed to investigate optical properties of graphene in magnetic and electric fields. The $2p_z$-dominated Hamiltonian is built from the subspace spanned by a lot of tight-binding functions in an enlarged unit cell. Its dimension is determined by the commensurate period of the geometric structure and external field. This model can be utilized in cases where many kinds of external fields are applied, e.g. uniform/modulated magnetic fields [124, 125, 150], modulated electric fields and composite fields [151]. This is suitable for multi-layer graphene [126, 128, 143, 144] and bulk graphite [255, 256] with arbitrary stacking configurations. The most important interlayer atomic interactions and external fields are simultaneously taken into account without the need to treat either of them as a perturbation term. The Hamiltonian matrix for determining the magneto-electric properties becomes too large for the experimental field strengths. By using an exact diagonalization method, the eigenvalues and eigenfunctions of the Hamiltonian matrix are efficiently solved. The developed model can be further used to understand other essential material properties by combining the single- and many-particle theories, such as optical spectra and electronic excitations [158, 159, 162, 175, 176, 190–196]. On the other hand, the effective mass approximation is frequently adopted to comprehend the electronic and optical properties. The low-lying energy bands are expanded about the high-symmetry point, and then magnetic quantization is done for the approximate properties. The magneto-optical absorption spectra are calculated for the AA- [127] and AB-stacked graphene [165]. Detailed comparisons of these two methods will be made in later discussions.

This book broadly covers the fields related to multilayer graphene systems in the forms of AA, AB, ABC, AAB and sliding configurations, and discusses many of the factors which affect their magneto-electronic properties. The absorption spectra of multilayer graphene systems are discussed in detail under various stacking configurations, and their competition and cooperation with the applied electric field and magnetic field are also discussed. The main spectral characteristics focus on the prominent structures, number, intensity and frequency of absorption peaks. In chapter 2, we introduce the Peierls tight-binding model under electric and magnetic fields. The optical response, used to describe the spectral properties, is calculated based on Fermi's golden rule. We start with monolayer graphene in chapter 3, and subsequently discuss AA-stacked trilayer graphene. Symmetry-preserved multi-Dirac-cone structures are responsible for the optical absorption spectra. Intra-Dirac-cone excitations comprise the whole frequency region of the spectra. Chapters 4 and 5 are devoted to AB- and ABC-stacked trilayer graphene. The excitations of the massless and massive Dirac quasi-particles dominate the former system, while the vertical transitions between any two inter- or intragroup LLs are permitted for the latter. In particular, the phenomenon of LL anticrossings in the energy spectrum is discussed and explained using a clear physical picture, especially for the energy width of the sombrero-shaped subbands. This leads to the corresponding non-monotonic dependence of the spectral frequency and intensity on the magnetic field. In chapters 6 and 7, we discuss the optical response under the effects of broken inversion symmetry in AAB-stacked trilayer and sliding bilayer graphene. The low-symmetry stacking configurations trigger more significant LL anticrossings and cause distinct features for the K- and K′-splitting LLs.

At the beginning of each chapter, we present the Peierls Hamiltonian and the spectral function with respect to each system. We focus on the mechanism of the optical excitations responsible for the spectral characteristics based on the LL spectra and wavefunctions. The electric-field-enriched optical properties are also discussed. Optical absorption experiments are available for the investigation of the layer and stacking dependence of the magneto-electronic properties. Meanwhile, a detailed comparison of these systems is also presented in chapter 7. Finally, chapter 8 contains the concluding remarks.

References

[1] Novoselov K S, Geim A K, Morozov S V, Jiang D, Zhang Y and Dubonos S V *et al* 2004 Electric field effect in atomically thin carbon films *Science* **306** 666–9

[2] Novoselov K S, Jiang D, Schedin F, Booth T J, Khotkevich V V and Morozov S V *et al* 2005 Two-dimensional atomic crystals *Proc. Natl Acad. Sci. USA.* **102** 10451–3

[3] Lee J K, Lee S C, Ahn J P, Kim S C, Wilson J I B and John P 2008 The growth of AA graphite on (1 1 1) diamond *J. Chem. Phys.* **129** 234709

[4] Liu Z, Suenaga K, Harris P J F and Iijima S 2009 Open and closed edges of graphene layers *Phys. Rev. Lett.* **102** 015501

[5] Borysiuk J, Soltys J and Piechota J 2011 Stacking sequence dependence of graphene layers on SiC (000-1)—experimental and theoretical investigation *J. Appl. Phys.* **109** 093523

[6] Cherkez V, Trambly de Laissardiere G, Mallet P and Veuillen J Y 2015 Van Hove singularities in doped twisted graphene bilayers studied by scanning tunneling spectroscopy *Phys. Rev.* B **91** 155428

[7] Yan W, Liu M, Dou R F, Meng L, Feng L and Chu Z D *et al* 2012 Angle-dependent van Hove singularities in a slightly twisted graphene bilayer *Phys. Rev. Lett.* **109** 126801

[8] Lauffer P, Emtsev K V, Graupner R, Seyller Th, Ley L and Reshanov S A *et al* 2008 Atomic and electronic structure of few-layer graphene on SiC(0001) studied with scanning tunneling microscopy and spectroscopy *Phys. Rev.* B **77** 155426
[9] Hattendorf S, Georgi A, Liebmann M and Morgenstern M 2013 Networks of ABA and ABC stacked graphene on mica observed by scanning tunneling microscopy *Surf. Sci.* **610** 53–8
[10] Norimatsu W and Kusunoki M 2010 Selective formation of ABC-stacked graphene layers on SiC(0001) *Phys. Rev.* B **81** 161410
[11] Warner J H, Mukai M and KirklandA I 2012 Atomic structure of ABC rhombohedral stacked trilayer graphene *ACS Nano.* **6** 5680–6
[12] Rong Z Y and Kuiper P 1993 Electronic effects in scanning tunneling microscopy: moiré pattern on a graphite surface *Phys. Rev.* B **48** 17427
[13] Campanera J M, Savini G, Suarez-Martinez I and Heggie M I 2007 Density functional calculations on the intricacies of moiré patterns on graphite *Phys. Rev.* B **75** 235449
[14] Biedermann L B, Bolen M L, Capano M A, Zemlyanov D and Reifenberger R G 2009 Insights into few-layer epitaxial graphene growth on 4H-SiC($000\bar{1}$) substrates from STM studies *Phys. Rev.* B **79** 125411
[15] Bernal J D 1924 The structure of graphite *Proc. R. Soc. London, Ser.* A **106** 749
[16] Lipson H and Stokes A R 1942 The structure of graphite *Proc. R. Sot. London, Ser.* A **181** 101
[17] Charlier J C, Gonze X and Michenaud J P 1994 First-principles study of the stacking effect on the electronic properties of graphite(s) *Carbon* **32** 289–99
[18] Charlier J C, Michenaud J P and Ph Lambin 1992 Tight-binding density of electronic states of pregraphitic carbon *Phys. Rev.* B **46** 4540–3
[19] Shallcross S, Sharma S, Kandelaki E and Pankratov O A 2010 Electronic structure of turbostratic graphene *Phys. Rev.* B **81** 165105
[20] Bracamonte M V, Lacconi G I, Urreta S E and Foa Torres L E F 2014 On the nature of defects in liquid-phase exfoliated graphene *J. Phys. Chem.* C **118** 15455–9
[21] Stankovich S, Dikin D A, Piner R D, KohlhaasK A, Kleinhammes A and Jia Y *et al* 2007 Synthesis of graphene-based nanosheets via chemical reduction of exfoliated graphite oxide *Carbon* **45** 1558–65
[22] Song P, Zhang X, Sun M, Cui X and Lin Y 2012 Synthesis of graphene nanosheets via oxalic acid-induced chemical reduction of exfoliated graphite oxide *RSC Adv.* **2** 1168–73
[23] Guo H L, Wang X F, Qian Q Y, Wang F B and Xia X H 2009 A green approach to the synthesis of graphene nanosheets *ACS Nano.* **3** 2653–9
[24] Dou L T, Cui F, Yu Y, Khanarian G, Eaton S W and Yang Q *et al* 2016 Solution-processed copper/reduced-graphene-oxide core/shell nanowire transparent conductors *ACS Nano.* **10** 2600–6
[25] Tang L, Wang Y, Li Y, Feng H, Lu J and Li J 2009 Preparation, structure, and electrochemical properties of reduced graphene sheet films *Adv. Funct. Mater.* **19** 2782–59
[26] Anis A, Mohammad A A, Hussain A and Ahmed S E 2016 Reduced graphene oxide thin film on conductive substrates by bipolar electrochemistry *Sci. Rep.* **6** 21282
[27] Sundaram R S, Gomez-Navarro C, Balasubramanian K, Burghard M, Marko K K and Klaus K 2008 Electrochemical modification of graphene *Adv. Mater.* **20** 3050–3
[28] Reina A, Jia X, Ho J, Nezich D, Son H and Bulovic V *et al* 2009 Large area, few-layer graphene films on arbitrary substrates by chemical vapor deposition *Nano Lett.* **9** 30–5
[29] Liu L, Zhou H, Cheng R, Yu W J, Liu Y and Chen Y *et al* 2012 High-yield chemical vapor deposition growth of high-quality large-area AB-stacked bilayer graphene *ACS Nano.* **6** 8241–9

[30] Gomez T, Florez E, Rodriguez J A and Illas F 2011 Reactivity of transition metals (Pd, Pt, Cu, Ag, Au) toward molecular hydrogen dissociation: extended surfaces versus particles supported on TiC(001) or small is not always better and large is not always bad *J. Phys. Chem.* C **115** 11666–72

[31] Zhang X, Li H and Ding F 2014 Self-assembly of carbon atoms on transition metal surfaces—chemical vapor deposition growth mechanism of graphene *Adv. Mater.* **26** 5488–95

[32] Zhou Z, Habenicht B F and Guo Q 2013 Graphene Moiré structure grown on a pseudomorphic metal overlayer supported on Ru(0001) *Surf. Sci.* **611** 67–73

[33] Kim K S, Zhao Y, Jang H, Lee S Y, Kim J M and Kim K S *et al* 2009 Large-scale pattern growth of graphene films for stretchable transparent electrodes *Nature* **457** 706–10

[34] Li X, Cai W, An J, Kim S, Nah J and Yang D *et al* 2009 Large-area synthesis of high-quality and uniform graphene films on copper foils *Science* **324** 1312–4

[35] Zhou H, Yu W J, Liu L, Cheng R, Chen Y and Huang X *et al* 2013 Chemical vapour deposition growth of large single crystals of monolayer and bilayer graphene *Nat. Commun.* **4** 2096

[36] Su C Y, Lu A Y, Wu C Y, Li Y T, Liu K K and Zhang W *et al* 2011 Direct formation of wafer scale graphene thin layers on insulating substrates by chemical vapor deposition *Nano Lett.* **11** 3612–6

[37] Bae S, Kim H, Lee Y, Xu X, Park J S and Zheng Y *et al* 2010 Roll-to-roll production of 30-inch graphene films for transparent electrodes *Nat. Nanotechnol.* **5** 574–8

[38] Vijayaraghavan R K, Gaman C, Jose B, Mccoy A P, Cafolla T and McNally P J *et al* 2016 Pulsed-plasma physical vapor deposition approach toward the facile synthesis of multilayer and monolayer graphene for anticoagulation applications *ACS Appl. Mater. Interfaces* **8** 4878–86

[39] Wu X Y, Zhong G F, D'Arsie L, Sugime H, Esconjauregui S and Robertson A W *et al* 2016 Growth of continuous monolayer graphene with millimeter-sized domains using industrially safe conditions *Sci. Rep.* **6** 21152

[40] Bosca A, Pedros J, Martinez J, Palacios T and Calle F 2016 Automatic graphene transfer system for improved material quality and efficiency *Sci. Rep.* **6** 21676

[41] Engel M, Steiner M, Lombardo A, Ferrari A C, v. Lohneysen H and Avouris P *et al* 2012 Light matter interaction in a micro-cavity controlled graphene transistor *Nat. Commun.* **3** 906–11

[42] Koppens F H L, Mueller T, Avouris Ph, Ferrari A C, Vitiello M S and Polini M 2014 Photodetectors based on graphene, other two-dimensional materials and hybrid systems *Nat. Nanotechnol.* **9** 780–93

[43] Bonaccorso F, Sun Z, Hasan T and Ferrari A C 2010 Graphene photonics and optoelectronics *Nat. Photonics* **4** 611–22

[44] Khan F, Baek S H and Kim J H 2016 One-step and controllable bipolar doping of reduced graphene oxide using TMAH as reducing agent and doping source for field effect transistors *Carbon* **100** 608–16

[45] Kumar A, Tyagi P, Dagar J and Srivastava R 2016 Tunable field effect properties in solid state and flexible graphene electronics on composite high-low k dielectric *Carbon* **99** 579–84

[46] Tseng S F, Haiso W T, Cheng P Y, Chung C K, Lin Y S and Chien S C *et al* 2016 Graphene-based chips fabricated by ultraviolet laser patterning for an electrochemical impedance spectroscopy *Sens. Actuators B Chem.* **226** 342–8

[47] Liu J B, Li P J, Chen Y F, Song X B, Mao Q and Wu Y *et al* 2016 Flexible terahertz modulator based on coplanar-gate graphene field-effect transistor structure *Optics Lett.* **41** 816–9

[48] Shown I, Hsu H C, Chang Y, Lin C H, Roy P K and Ganguly A *et al* 2014 Highly efficient visible light photocatalytic reduction of CO_2 to hydrocarbon fuels by Cu-nanoparticle decorated graphene oxide *Nano Lett.* **14** 6097–103

[49] Hsu H C, Shown I, Wei H Y, Chang Y C, Du H Y and Lin Y G *et al* 2013 Graphene oxide as a promising photocatalyst for CO_2 to methanol conversion *Nanoscale* **5** 262–8

[50] Idier J, Neri W, Labrugere C, Ly I, Poulin P and Backov R 2016 Modified silver nanowire transparent electrodes with exceptional stability against oxidation *Nanotechnology* **27** 105705

[51] Koppens F H L, Chang D E and de Abajo F J G 2011 Graphene plasmonics: a platform for strong light-matter interactions *Nano Lett.* **11** 3370–7

[52] Matyba P, Yamaguchi H, Eda G, Chhowalla M, Edman L and Robinson N D 2010 Graphene and mobile ions: the key to all-plastic, solution-processed light-emitting devices *ACS Nano* **4** 637–42

[53] Kusumawati Y, Koussi-Daoud S and Pauporte T 2016 TiO_2/graphene nanocomposite layers for improving the performances of dye-sensitized solar cells using a cobalt redox shuttle *J. Photochem. Photobiol.* **329** 54–60

[54] Xia F N, Mueller T, Lin Y M, Valdes-Garcia A and Avouris P 2009 Ultrafast graphene photodetector *Nat. Nanotechnol.* **4** 839–43

[55] Mueller T, Xia F and Avouris P 2010 Graphene photodetectors for high-speed optical communications *Nat. Photon.* **4** 297–301

[56] Liu M, Yin X B, Ulin-Avila E, Geng B S, Zentgraf T and Ju L *et al* 2011 A graphene-based broadband optical modulator *Nature* **474** 64–7

[57] Xu P, Yang Y, Qi D, Barber S D, Schoelz J K and Ackerman M L *et al* 2012 Electronic transition from graphite to graphene via controlled movement of the top layer with scanning tunneling microscopy *Phys. Rev.* B **86** 085428

[58] Rong Z Y 1994 Extended modifications of electronic structures caused by defects: scanning tunneling microscopy of graphite *Phys. Rev.* B **50** 1839–43

[59] Pong W T, Bendall J and Durkan C 2007 Observation and investigation of graphite superlattice boundaries by scanning tunneling microscopy *Surf. Sci.* **601** 498

[60] Campanera J M, Savini G, Suarez-Martinez I and Heggie M I 2007 Density functional calculations on the intricacies of moiré patterns on graphite *Phys. Rev.* B **75** 235449

[61] Kazemi A S, Crampin S and Ilie A 2013 Stacking-dependent superstructures at stepped armchair interfaces of bilayer/trilayer graphene *Appl. Phys. Lett.* **102** 163111

[62] Biedermann L B, Bolen M L, Capano M A, Zemlyanov D and Reifenberger R G 2009 Insights into few-layer epitaxial graphene growth on 4H-SiC(000$\bar{1}$) substrates from STM studies *Phys. Rev.* B **79** 125411

[63] Que Y, Xiao W, Chen H, Wang D, Du S and Gao H J 2015 Stacking-dependent electronic property of trilayer graphene epitaxially grown on Ru(0001) *Appl. Phys. Lett.* **107** 263101

[64] Do T N, Lin C Y, Lin Y P, Shih P H and Lin M F 2015 Configuration-enriched magneto-electronic spectra of AAB-stacked trilayer graphene *Carbon* **94** 619–32

[65] Do T N, Shih P H, Chang C P, Lin C Y and Lin M F 2015 Rich magneto-absorption spectra in AAB-stacked trilayer graphene *Phys. Chem. Chem. Phys.* **18** 17597–605

[66] Huang Y K, Chen S C, Ho Y H, Lin C Y and Lin M F 2014 Feature-rich magnetic quantization in sliding bilayer graphenes *Sci. Rep.* **4** 7509

[67] Kindermann M and Mele E J 2011 Landau quantization in twisted bilayer graphene: the Dirac comb *Phys. Rev.* B **84** 161406

[68] Ohta T, Bostwick A and McChesney J L 2007 Interlayer interaction and electronic screening in multilayer graphene investigated with angle-resolved photoemission spectroscopy *Phys. Rev. Lett.* **98** 206802

[69] Moreau E, Godey S, Wallart X, Razado-Colambo I, Avila J and Asensio M C *et al* 2013 High-resolution angle-resolved photoemission spectroscopy study of monolayer and bilayer graphene on the C-face of SiC *Phys. Rev.* B **88** 075406

[70] Coletti C, Forti S, Principi A, Emtsev K V, Zakharov A A and Daniels K M *et al* 2013 Revealing the electronic band structure of trilayer graphene on SiC: an angle-resolved photoemission study *Phys. Rev.* B **88** 155439

[71] Pletikosić I, Kralj M, Pervan P, Brako R, Coraux J and N'Diaye A T *et al* 2009 Dirac cones and minigaps for graphene on Ir(111) *Phys. Rev. Lett.* **102** 056808

[72] Bostwick A, Ohta T, Seyller T, Horn K and Rotenberg E 2007 Quasiparticle dynamics in graphene *Nat. Phys.* **3** 36–40

[73] Emtsev K V, Speck F, Seyller T, Ley L and Riley J D 2008 Interaction, growth, and ordering of epitaxial graphene on SiC0001 surfaces: a comparative photoelectron spectroscopy study *Phys. Rev.* B **77** 155303

[74] Sprinkle M, Siegel D, Hu Y, Hicks J, Tejeda A and Taleb-Ibrahimi A *et al* 2009 First direct observation of a nearly ideal graphene band structure *Phys. Rev. Lett.* **103** 226803

[75] Zhang X, Han W P, Qiao X F, Tan Q H, Wang Y F and Zhang J *et al* 2016 Raman characterization of AB- and ABC-stacked few-layer graphene by interlayer shear modes *Carbon* **99** 118–22

[76] Raghavan S, Denig T J, Nelson T C, Chaudhari S and Stinespring C D 2016 Characterization of graphene-on-insulator films formed using plasma based surface chemistry *Carbon* **99** 212–21

[77] Wallace P R 1947 The band theory of graphite *Phys. Rev.* **71** 622–34

[78] Tsai S J, Chiu Y H, Ho Y H and Lin M F 2012 Gate-voltage-dependent Landau levels in AA-stacked bilayer graphene *Chem. Phys. Lett.* **550** 104–10

[79] Wang D and Jin G 2013 Tunneling magnetoresistance tuned by a vertical electric field in an AA-stacked graphene bilayer with double magnetic barriers *J. Appl. Phys.* **114** 233701

[80] Lobato I and Partoens B 2011 Multiple Dirac particles in AA-stacked graphite and multilayers of graphene *Phys. Rev.* B **83** 165429

[81] Lu C L, Chang C P, Huang Y C, Lu JM, Hwang C C and Lin M F 2006 Low-energy electronic properties of the AB-stacked few-layer graphites *J. Phys.: Condens. Matter* **18** 5849–59

[82] Gruneis A, Attaccalite C, Wirtz L, Shiozawa H, Saito R and Pichler T *et al* 2008 Tight-binding description of the quasiparticle dispersion of graphite and few-layer graphene *Phys. Rev.* B **78** 205425

[83] McCann E and Koshino M 2013 The electronic properties of bilayer graphene *Rep. Prog. Phys.* **76** 056503

[84] Partoens B and Peeters F M 2007 Normal and Dirac fermions in graphene multilayers: tight-binding description of the electronic structure *Phys. Rev.* B **75** 193402

[85] Luican A, Li G H, Reina A, Kong J, Nair R R and Novoselov K S *et al* 2011 Single-layer behavior and its breakdown in twisted graphene layers *Phys. Rev. Lett.* **106** 126802

[86] Kim K S, Walter A L, Moreschini L, Seyller T, Horn K and Rotenberg E *et al* 2013 Coexisting massive and massless Dirac fermions in symmetry-broken bilayer graphene *Nat. Mater.* **12** 887–92

[87] Li G H, Luican A, Lopes dos Santos J M B, Castro Neto A H, Reina A and Kong J *et al* 2010 Observation of van Hove singularities in twisted graphene layers *Nat. Phys.* **6** 109–13
[88] Goto H, Uesugi E, Eguchi R and Kubozono Y 2013 Parity effects in few-layer graphene *Nano Lett.* **3** 5153–8
[89] Latil S, Meunier V and Henrard L 2007 Massless fermions in multilayer graphitic systems with misoriented layers: *ab initio* calculations and experimental fingerprints *Phys. Rev.* B **76** 201402 (R)
[90] Koshino M and McCann E 2010 Parity and valley degeneracy in multilayer graphene *Phys. Rev.* B **81** 115315
[91] Maka K F, Sfeirb M Y, Misewichb J A and Heinza T F 2010 The evolution of electronic structure in few-layer graphene revealed by optical spectroscopy *Proc. Natl Acad. Sci. USA.* **107** 14999–15004
[92] Yan X Z and Ting C S 2015 Current orderings of interacting electrons in bilayer graphene *Phys. Rev.* B **92** 075442
[93] Castro Neto A H, Guinea F, Peres N M R, Novoselov K S and Geim A K 2009 The electronic properties of graphene *Rev. Mod. Phys.* **81** 109–162
[94] Latil S and Henrard L 2006 Charge carriers in few-layer graphene films *Phys. Rev. Lett.* **97** 036803
[95] Zhang F, Sahu B, Min H and MacDonald A H 2010 Band structure of ABC-stacked graphene trilayers *Phys. Rev.* B **82** 035409
[96] Xiao R, Tasnadi F, Koepernik K, Venderbos J W F, Richter M and Taut M 2011 Density functional investigation of rhombohedral stacks of graphene: topological surface states, nonlinear dielectric response, and bulk limit *Phys. Rev.* B **84** 165404
[97] Guinea F, Castro Neto A H and Peres N M R 2006 Electronic states and Landau levels in graphene stacks *Phys. Rev.* B **73** 245426
[98] Wang H, Gao J H and Zhang F C 2013 Flat band electrons and interactions in rhombohedral trilayer graphene *Phys. Rev.* B **87** 155116
[99] Kopnin N B, Heikkila T T and Volovik G E 2011 High-temperature surface superconductivity in topological flat-band systems *Phys. Rev.* B **83** 220503
[100] Lu C L, Chang C P, Huang Y C, Chen R B and Lin M L 2006 Influence of an electric field on the optical properties of few-layer graphene with AB stacking *Phys. Rev.* B **73** 144427
[101] Lu C L, Chang C P, Huang Y C, Ho J H, Hwang C C and Lin M F 2007 Electronic properties of AA- and ABC-stacked few-layer graphites *J. Phys. Soc. Jap.* **76** 024701
[102] McCann E 2006 Asymmetry gap in the electronic band structure of bilayer graphene *Phys. Rev.* B **74** 161403
[103] Min H, Sahu B, Banerjee S K and MacDonald A H 2007 *Ab initio* theory of gate induced gaps in graphene bilayers *Phys. Rev.* B **75** 155115
[104] Kumar S B and Guo J 2011 Multilayer graphene under vertical electric field *App. Phy. Lett.* **98** 222101
[105] Wu B R 2011 Field modulation of the electronic structure of trilayer graphene *App. Phy. Lett.* **98** 263107
[106] Avetisyan A A, Partoens B and Peeters F M 2009 Electric field tuning of the band gap in graphene multilayers *Phys. Rev.* B **79** 035421
[107] Castro E V, Novoselov K S, Morozov S V, Peres N M R, Lopes dos Santos J M B and Nilsson J *et al* 2007 Biased bilayer graphene: semiconductor with a gap tunable by the electric field effect *Phys. Rev. Lett.* **99** 216802

[108] Yankowitz M, Wang F L, Lau C N and LeRoy B J 2013 Local spectroscopy of the electrically tunable band gap in trilayer graphene *Phys. Rev.* B **87** 165102

[109] Zhang Y, Tang T T, Girit C, Hao Z, Martin M C and Zettl A *et al* 2009 Direct observation of a widely tunable bandgap in bilayer graphene *Nature* **459** 820–3

[110] Lui C H, Li Z, Mak K F, Cappelluti E and Heinz T F 2011 Observation of an electrically tunable band gap in trilayer graphene *Nat. Phys.* **7** 944–7

[111] Taisuke O, Aaron B and Thomas S 2006 Controlling the electronic structure of bilayer graphene *Science* **313** 951–4

[112] Bao W, Jing L, Velasco J, Lee Y, Liu G and Tran D *et al* 2011 Stacking-dependent band gap and quantum transport in trilayer graphene *Nat. Phys.* **7** 948–52

[113] Wang F, Zhang Y, Tian C, Girit C, Zettl A and Crommie M *et al* 2008 Gate-variable optical transitions in graphene *Science* **320** 206–9

[114] Craciun M F, Russo S, Yamamoto M, Oostinga J B, Morpurgo A F and Tarucha S 2009 Trilayer graphene is a semimetal with a gate-tunable band overlap *Nat. Nanotechnol.* **4** 383–8

[115] Zhang W, Lin C T, Liu K K, Tite T, Su C Y and Chang C H *et al* 2011 Opening electrical band gap of bilayer graphene with molecular doping *ACS Nano.* **5** 7517–24

[116] Lee C, Kim J, Kim S, Chang Y J, Kim K S and Hong B *et al* 2016 Strong hole-doping and robust resistance-decrease in proton-irradiated graphene *Sci. Rep.* **6** 21311

[117] Liu H, Liu Y and Zhu D 2011 Chemical doping of graphene *J. Mater. Chem.* **21** 3335–45

[118] Wei D, Liu Y, Wang Y, Zhang H, Huang L and Yu G 2009 Synthesis of N-doped graphene by chemical vapor deposition and its electrical properties *Nano Lett.* **9** 1752–8

[119] Nakada K and Ishii A 2011 Migration of adatom adsorption on graphene using DFT calculation *Solid State Commun.* **151** 13–6

[120] Rana M, Arora G and Gautam U K 2015 N- and S-doped high surface area carbon derived from soya chunks as scalable and efficient electrocatalysts for oxygen reduction *Sci. Technol. Adv. Mat.* **16** 014803

[121] Lin Y P, Ksari Y, Aubel D, Hajjar-Garreau S, Borvon G and Spiegel Y *et al* 2016 Efficient and low-damage nitrogen doping of graphene via plasma-based methods *Carbon* **100** 337–44

[122] Arjmand M, Chizari K, Krause B, Potschke P and Sundararaj U 2016 Effect of synthesis catalyst on structure of nitrogen-doped carbon nanotubes and electrical conductivity and electromagnetic interference shielding of their polymeric nanocomposites *Carbon* **98** 358–72

[123] Raquet B, Avriller R, Lassagne B, Nanot S, Escoffier W and Broto JM *et al* 2008 Onset of Landau-level formation in carbon-nanotube-based electronic Fabry-Pérot resonators *Phys. Rev. Lett.* **101** 046803

[124] Chang C P, Lu C L, Shyu F L, Chen R B, Fang Y K and Lin M F 2004 Magnetoelectronic properties of a graphite sheet *Carbon* **42** 2975–80

[125] Ho J H, Lai Y H, Chiu Y H and Lin M F 2008 Landau levels in graphene *Physica* E **40** 1722–5

[126] Lin C Y, Wu J Y, Ou Y J, Chiu Y H and Lin M F 2015 Magneto-electronic properties of multilayer graphenes *Phys. Chem. Chem. Phys.* **17** 26008–35

[127] Chang C P 2011 Exact solution of the spectrum and magneto-optics of multilayer hexagonal graphene *J. Appl. Phys.* **110** 013725

[128] Lai Y H, Ho J H, Chang C P and Lin M F 2008 Magnetoelectronic properties of bilayer bernal graphene *Phys. Rev.* B **77** 085426

[129] McCann E and Fal'ko V I 2006 Landau-level degeneracy and quantum Hall effect in a graphite bilayer *Phys. Rev. Lett.* **96** 086805

[130] Min H and MacDonald A H 2008 Chiral decomposition in the electronic structure of graphene multilayers *Phys. Rev.* B **77** 155416

[131] Koshino M and McCann E 2011 Landau level spectra and the quantum Hall effect of multilayer graphene *Phys. Rev.* B **83** 165443

[132] Min H and MacDonald A H 2008 Electronic structure of multilayer graphene *Prog. Theor. Phys. Supplement* **176** 227–52

[133] Aoki M and Amawash H 2007 Dependence of band structures on stacking and field in layered graphene *Solid State Commun.* **142** 123–7

[134] Avetisyan A A, Partoens B and Peeters F M 2010 Stacking order dependent electric field tuning of the band gap in graphene multilayers *Phys. Rev.* B **81** 115432

[135] Koshino M and McCann E 2009 Gate-induced interlayer asymmetry in ABA-stacked trilayer graphene *Phys. Rev.* B **79** 125443

[136] Kim D S, Kwon H, Nikitin A Y, Ahn S, Martin-Moreno L and Garcia-Vidal F J *et al* 2015 Stacking structures of few-layer graphene revealed by phase-sensitive infrared nanoscopy *ACS Nano* **9** 6765–73

[137] Cadden-Zimansky P, Jiang Z, Li Z Q, Tung L C, Schwartz M E and Takita M *et al* 2010 Interaction-induced shift of the cyclotron resonance of graphene using infrared spectroscopy *Phys. Rev. Lett.* **104** 067404

[138] Faugeras C, Amado M, Kossacki P, Orlita M, Kuhne M and Nicolet A A L *et al* 2011 Magneto-Raman scattering of graphene on graphite: electronic and phonon excitations *Phys. Rev. Lett.* **107** 036807

[139] Zhao H M, Lin Y C, Yeh C H, Tian H, Chen Y C and Xie D *et al* 2014 Growth and raman spectra of single-crystal trilayer graphene with different stacking orientations *ACS Nano.* **8** 10766–73

[140] Cong C X, Jung J, Cao B C, Qiu C Y, Shen X N and Ferreira A *et al* 2015 Magnetic oscillation of optical phonon in ABA- and ABC-stacked trilayer graphene *Phys. Rev.* B **91** 235403

[141] Sena S H R, Pereira J M, Peeters F M and Farias G A 2011 Landau levels in asymmetric graphene trilayers *Phys. Rev.* B **84** 205448

[142] Yuan S, Roldan R and Katsnelson M I 2011 Landau level spectrum of ABA- and ABC-stacked trilayer graphene *Phys. Rev.* B **84** 125455

[143] Ho C H, Tsai S J, Chen R B, Chiu Y H and Lin M F 2011 Low-energy Landau level spectrum in ABC-stacked trilayer graphene *J. Nanosci. Nanotechnol.* **11** 4938–47

[144] Ho C H, Ho Y H, Chiu Y H, Chen Y N and Lin M F 2011 Magneto-electronic properties of rhombohedral trilayer graphene: Peierls tight-binding model *Ann. Phys.* **326** 721–39

[145] Koshino M and McCann E 2009 Trigonal warping and Berry's phase $N\pi$ in ABC-stacked multilayer graphene *Phy. Rev.* B **80** 165409

[146] Lin C Y, Wu J Y, Chiu Y H, Chang C P and Lin M F 2014 Stacking-dependent magnetoelectronic properties in multilayer graphene *Phys. Rev.* B **90** 205434

[147] Lin Y P, Wang J, Lu J M, Lin C Y and Lin M F 2014 Energy spectra of ABC-stacked trilayer graphene in magnetic and electric fields *RSC Adv.* **4** 56552–60

[148] Tran N T T, Lin S Y, Glukhova O E and Lin M F 2015 Configuration-induced rich electronic properties of bilayer graphene *J Phys. Chem.* C **119** 10623–30

[149] Inoue M 1962 Landau levels and cyclotron resonance in graphite *J. Phys. Soc. Jpn* **17** 808

[150] Ou Y C, Sheu J K, Chiu Y H, Chen R B and Lin M F 2011 Influence of modulated fields on the Landau level properties of graphene *Phys. Rev.* B **83** 195405

[151] Chiu Y H, Ou Y C, Liao Y Y and Lin M F 2010 Optical-absorption spectra of single-layer graphene in a periodic magnetic field *J. Vac. Sci. Technol.* B **28** 386–90
[152] Ou Y C, Chiu Y H, Yang P H and Lin M F 2014 The selection rule of graphene in a composite magnetic field *Opti. Express* **22** 7473–91
[153] Chiu C W, Chen S C, Huang Y C, Shyu F L and Lin M F 2013 Critical optical properties of AA-stacked multilayer graphenes *Appl. Phys. Lett.* **103** 041907
[154] Lee S H, Ho Y H, Chiu C W and Lin M F 2010 Optical properties of deformed few-layer graphenes with AB stacking *J. Appl. Phys.* **108** 043509
[155] Kadi F and Malic E 2014 Optical properties of bernal-stacked bilayer graphene: a theoretical study *Phys. Rev.* B **89** 045419
[156] Koshino M 2013 Stacking-dependent optical absorption in multilayer graphene *New J. Phys.* **15** 015010
[157] Chiu C W, Ho Y H, Shyu F L and Lin M F 2014 Layer-enriched optical spectra of AB-stacked multilayer graphene *App. Phys. Express* **7** 115102
[158] Ho Y H, Chiu Y H, Lin D H, Chang C P and Lin M F 2010 Magneto-optical selection rules in bilayer Bernal graphene *ACS Nano* **4** 1465–72
[159] Abergel D S L and Fal'ko V I 2007 Optical and magneto-optical far-infrared properties of bilayer graphene *Phys. Rev.* B **75** 155430
[160] Chiu C W, Huang Y C, Shyu F L and Lin M F 2012 Optical absorption spectra in ABC-stacked graphene superlattice *Synthetic Met.* **162** 800–4
[161] Yan J A, Ruan W Y and Chou M Y 2011 Enhanced optical conductivity induced by surface states in ABC-stacked few-layer graphene *Phys. Rev.* B **83** 245418
[162] Lin C Y, Wu J Y, Chang C P and Lin M F 2014 Magneto-optical selection rules of curved graphene nanoribbons and carbon nanotubes *Carbon* **69** 151–61
[163] Gusynin V P and Sharapov S G 2006 Transport of Dirac quasiparticles in graphene: Hall and optical conductivities *Phys. Rev.* B **73** 245411
[164] Gusynin V P, Sharapov S G and Carbotte J P 2007 Anomalous absorption line in the magneto-optical response of graphene *Phys. Rev. Lett.* **98** 157402
[165] Koshino M and Ando T 2008 Magneto-optical properties of multilayer graphene *Phy. Rev.* B **77** 115313
[166] Li Z Q, Henriksen E A, Jiang Z, Hao Z, Martin M C and Kim P *et al* 2008 Dirac charge dynamics in graphene by infrared spectroscopy *Nat. Phys.* **4** 532–5
[167] Zhang L M, Li Z Q, Basov D N, Fogler M M, Hao Z and Martin M C 2008 Determination of the electronic structure of bilayer graphene from infrared spectroscopy *Phys. Rev.* B **78** 235408
[168] Kuzmenko A B, van Heumen E, van der Marel D, Lerch P, Blake P and Novoselov K S *et al* 2009 Infrared spectroscopy of electronic bands in bilayer graphene *Phys. Rev.* B **79** 115441
[169] Faugeras C, Berciaud S, Leszczynski P, Henni Y, Nogajewski K and Orlita M *et al* 2015 Landau level spectroscopy of electron–electron interactions in graphene *Phys. Rev. Lett.* **114** 126804
[170] Orlita M, Faugeras C, Barra A L, Martinez G, Potemski M and Basko D M *et al* 2015 Infrared magneto-spectroscopy of two-dimensional and three-dimensional massless fermions: a comparison *J. Appl. Phys.* **117** 112803
[171] Berciaud S, Potemski M and Faugeras C 2014 Probing electronic excitations in mono- to pentalayer graphene by micro magneto-Raman spectroscopy *Nano Lett.* **14** 4548–53

[172] Faugeras C, Binder J, Nicolet A A L, Leszczynski P, Kossacki P and Wysmolek A *et al* 2014 A micro-magneto-Raman scattering study of graphene on a bulk graphite substrate *EPL* **108** 27011

[173] Li Z Q, Henriksen E A, Jiang Z, Hao Z, Martin M C and Kim P *et al* 2009 Band structure asymmetry of bilayer graphene revealed by infrared spectroscopy *Phys. Rev. Lett.* **102** 037403

[174] Mak K F, Shan J and Heinz T F 2010 Electronic structure of few-layer graphene: experimental demonstration of strong dependence on stacking sequence *Phys. Rev. Lett.* **104** 176404

[175] Lin Y P, Lin C Y, Ho Y H, Do T N and Lin M F 2015 Magneto-optical properties of ABC-stacked trilayer graphene *Phys. Chem. Chem. Phys.* **17** 15921–7

[176] Lin Y P, Lin C Y, Chang C P and Lin M F 2015 Electric-field-induced rich magneto-absorption spectra of ABC-stacked trilayer graphene *RSC Adv.* **5** 80410

[177] Jiang Z, Henriksen E A, Tung L C, Wang Y J, Schwartz M E and Han M Y *et al* 2007 Infrared spectroscopy of Landau levels of graphene *Phys. Rev. Lett.* **98** 197403

[178] Kuhne P, Darakchieva V, Yakimova R, Tedesco J D, Myers-Ward R L and Eddy C R *et al* 2013 Polarization selection rules for inter-Landau-level transitions in epitaxial graphene revealed by the infrared optical Hall effect *Phys. Rev. Lett.* **111** 077402

[179] Orlita M, Faugeras C, Plochocka P, Neugebauer P, Martinez G and Maude D K *et al* 2008 Approaching the Dirac point in high-mobility multilayer epitaxial graphene *Phys. Rev. Lett.* **101** 267601

[180] Plochocka P, Faugeras C, Orlita M, Sadowski M L, Martinez G and Potemski M *et al* 2008 High-energy limit of massless Dirac fermions in multilayer graphene using magneto-optical transmission spectroscopy *Phys. Rev. Lett.* **100** 087401

[181] Daboussi A, Mandhour L, Fuchs J N and Jaziri S 2014 Tunable zero-energy transmission resonances in shifted graphene bilayer *Phys. Rev.* B **89** 085426

[182] Pedersen J G and Pedersen T G 2011 Tight-binding study of the magneto-optical properties of gapped graphene *Phys. Rev.* B **84** 115424

[183] Ho Y H, Wu J Y, Chen R B and Lin M F 2010 Optical transitions between Landau levels: AA-stacked bilayer graphene *App. Phys. Lett.* **97** 101905

[184] Orlita M, Faugeras C, Borysiuk J, Baranowski J M, Strupiński W and Sprinkle M *et al* 2011 Magneto-optics of bilayer inclusions in multilayered epitaxial graphene on the carbon face of SiC *Phys. Rev.* B **83** 125302

[185] Orlita M, Faugeras C, Schneider JM, Martinez G, Maude D K and Potemski M 2009 Graphite from the viewpoint of Landau level spectroscopy: an effective graphene bilayer and monolayer *Phys. Rev. Lett.* **102** 166401

[186] Henriksen E A, Jiang Z, Tung L C, Schwartz M E, Takita M and Wang Y J *et al* 2008 Cyclotron resonance in bilayer graphene *Phys. Rev. Lett.* **100** 087403

[187] Ellis C T, Stier A V, Kim M H, Tischler J G, Glaser E R and Myers-Ward R L *et al* 2013 Magneto-optical fingerprints of distinct graphene multilayers using the giant infrared Kerr effect *Sci. Rep.* **3** 3143

[188] Tung L C, Cadden-Zimansky P, Qi J, Jiang Z and Smirnov D 2011 Measurement of graphite tight-binding parameters using high-field magnetoreflectance *Phys. Rev.* B **84** 153405

[189] Plochocka P, Solane P Y, Nicholas R J, Schneider J M, Piot B A and Maude D K *et al* 2012 Origin of electron-hole asymmetry in graphite and graphene *Phys. Rev.* B **85** 245410

[190] Ho J H, Lu C L, Hwang C C, Chang C P and Lin M F 2006 Coulomb excitations in AA- and AB-stacked bilayer graphites *Phys. Rev.* B **74** 085406
[191] Ho J H, Chang C P and Lin M F 2005 Electronic excitations of the multilayered graphite *Phys. Lett.* A **352** 446–50
[192] Lin M F, Chuang Y C and Wu J Y 2012 Electrically tunable plasma excitations in AA-stacked multilayer graphene *Phys. Rev.* B **86** 125434
[193] Wu J Y, Chen S C, Roslyak O, Gumbs G and Lin M F 2011 Plasma excitations in graphene: their spectral intensity and temperature dependence in magnetic field *ACS Nano* **5** 1026–32
[194] Lozovik Y E and Sokolik A A 2012 Influence of Landau level mixing on the properties of elementary excitations in graphene in strong magnetic field *Nano Res. Lett.* **7** 134
[195] Wu J Y, Gumbs G and Lin M F 2014 Combined effect of stacking and magnetic field on plasmon excitations in bilayer graphene *Phys. Rev.* B **89** 165407
[196] Wu J Y, Lin C Y, Gumbs G and Lin M F 2015 The effect of perpendicular electric field on temperature-induced plasmon excitations for intrinsic silicene *RSC Adv.* **5** 51912–8
[197] Yuan S, Roldan R and Katsnelson M I 2011 Excitation spectrum and high-energy plasmons in single-layer and multilayer graphene *Phys. Rev.* B **84** 035439
[198] Wang W, Kinaret J M and Apell S P 2012 Excitation of edge magnetoplasmons in semi-infinite graphene sheets: temperature effects *Phys. Rev.* B **85** 235444
[199] Roldan R, Fuchs J N and Goerbig M O 2009 Collective modes of doped graphene and a standard two-dimensional electron gas in a strong magnetic field: linear magnetoplasmons versus magnetoexcitons *Phys. Rev.* B **80** 085408
[200] Roldan R and Brey L 2013 Dielectric screening and plasmons in AA-stacked bilayer graphene *Phy. Rev.* B **88** 115420
[201] Gumbs G and Huang D 2007 Electronically modulated two-dimensional plasmons coupled to surface plasmon modes *Phys. Rev.* B **75** 115314
[202] Roslyak O, Gumbs G and Huang D 2011 Plasma excitations of dressed Dirac electrons in graphene layers *J. Appl. Phys.* **109** 113721
[203] Geim A K and Novoselov K S 2007 The rise of graphene *Nat. Mater.* **6** 183–91
[204] Novoselov K S, Geim A K, Morozov S V, Jiang D, Katsnelson M I and Grigorieva I V *et al* 2005 Two-dimensional gas of massless Dirac fermions in graphene *Nature* **438** 197–200
[205] Hao Y, Bharathi M S, Wang L, Liu Y, Chen H and Nie S *et al* 2013 The role of surface oxygen in the growth of large single-crystal graphene on copper *Science* **342** 720–3
[206] Chen J H, Jang C, Xiao S, Ishigami M and Fuhrer M S 2008 Intrinsic and extrinsic performance limits of graphene devices on SiO_2 *Nat. Nanotechnol.* **3** 206–9
[207] Morozov S V, Novoselov K S, Katsnelson M I, Schedin F, Elias D C and Jaszczak J A *et al* 2008 Giant intrinsic carrier mobilities in graphene and its bilayer *Phys. Rev. Lett.* **100** 016602
[208] Pallecchi E, Lafont F, Cavaliere V, Schopfer F, Mailly D and Poirier W *et al* 2014 High electron mobility in epitaxial graphene on 4H-SiC(0001) via post-growth annealing under hydrogen *Sci. Rep.* **4** 4558
[209] Schwierz F 2010 Graphene transistors *Nat. Nanotechnol.* **5** 487–96
[210] Gusynin V P and Sharapov S G 2005 Unconventional integer quantum Hall effect *Phys. Rev. Lett.* **95** 146801
[211] Zhang Y B, Tan Y W, Stormer H L and Kim P 2005 Experimental observation of the quantum Hall effect and Berry's phase in graphene *Nature* **438** 201–4

[212] Novoselov K S, Jiang Z, Zhang Y, Morozov S V, Stormer H L and Zeitler U *et al* 2007 Room-temperature quantum Hall effect in graphene *Science* **315** 1379–9

[213] Hsu Y F and Guo G Y 2010 Anomalous integer quantum Hall effect in AA-stacked bilayer graphene *Phys. Rev.* B **82** 165404

[214] Novoselov K S, McCann E, Morozov S V, Fal'ko V I, Katsnelson M I and Zeitler U *et al* 2006 Unconventional quantum Hall effect and Berry's phase of 2π in bilayer graphene *Nat. Phys.* **2** 177–80

[215] Oostinga J B, Heersche H B, Liu X, Morpurgo A F and Vandersypen L M K 2008 Gate-induced insulating state in bilayer graphene devices *Nat. Mater.* **7** 151–7

[216] Taychatanapat T, Watanabe K, Taniguchi T and Jarillo-Herrero P 2008 Quantum Hall effect and Landau-level crossing of Dirac fermions in trilayer graphene *Nat. Phys.* **7** 621–5

[217] van Elferen H J, Veligura A, Tombros N, Kurganova E V, van Wees B J and Maan J C *et al* 2013 Fine structure of the lowest Landau level in suspended trilayer graphene *Phys. Rev.* B **88** 121302

[218] Kumar A, Escoffier W, Poumirol J M, Faugeras C, Arovas D P and Fogler M M *et al* 2011 Integer quantum Hall effect in trilayer graphene *Phys. Rev. Lett.* **107** 126806

[219] Zhang L, Zhang Y, Camacho J, Khodas M and Zaliznyak I 2011 The experimental observation of quantum Hall effect of $l = 3$ chiral quasiparticles in trilayer graphene *Nat. Phys.* **7** 953–7

[220] Lee Y, Velasco J, Tran D, Zhang F, Bao W and Jing L *et al* 2013 Broken symmetry quantum Hall states in dual-gated ABA trilayer graphene *Nano Lett.* **13** 1627–31

[221] Henriksen E A, Nandi D and Eisenstein J P 2012 Quantum Hall effect and semimetallic behavior of dual-gated ABA-stacked trilayer graphene *Phys. Rev.* X **2** 011004

[222] Foa Torres L E F, Perez-Piskunow P M, Balseiro C A and Usaj G 2014 Multiterminal conductance of a Floquet topological insulator *Phys. Rev. Lett.* **113** 266801

[223] Ezawa M 2012 A topological insulator and helical zero mode in silicene under an inhomogeneous electric field *New J. Phys.* **14** 033003

[224] Ezawa M 2012 Valley-polarized metals and quantum anomalous Hall effect in silicene *Phys. Rev. Lett.* **109** 055502

[225] Tai Y Y, Choi H C, Ahmed T, Ting C S and Zhu J X 2015 Edge states and local electronic structure around an adsorbed impurity in a topological superconductor *Phys. Rev.* B **92** 174514

[226] Li M Y, Shi Y, Cheng C C, Lu L S, Lin Y C and Tang H L *et al* 2015 Epitaxial growth of a monolayer WSe_2-MoS_2 lateral p-n junction with an atomically sharp interface science **349** 524–58

[227] Ho Y H, Wang Y H and Chen H Y 2014 Magnetoelectronic and optical properties of a MoS_2 monolayer *Phys. Rev.* B **89** 155316

[228] Ho Y H, Chiu C W and Lin M F 2014 Magneto-optical spectra of transition metal dichalcogenides: a comparative study *Appl. Phys. Lett.* **105** 222411

[229] Tabert C J and Nicol E J 2013 Valley-spin polarization in the magneto-optical response of silicene and other similar 2D crystals *Phys. Rev. Lett.* **110** 197402

[230] Tabert C J and Nicol E J 2013 Magneto-optical conductivity of silicene and other buckled honeycomb lattices *Phys. Rev.* B **88** 085434

[231] Schafgans A A, Post K W, Taskin A A, Ando Y, Qi X L and Chapler B C *et al* 2012 Landau level spectroscopy of surface states in the topological insulator $Bi_{0.91}Sb_{0.09}$ via magneto-optics *Phys. Rev.* B **85** 195440

[232] LaForge A D, Frenzel A, Pursley B C, Lin T, Liu X F and Shi J *et al* 2010 Optical characterization of Bi_2Se_3 in a magnetic field: infrared evidence for magnetoelectric coupling in a topological insulator material *Phys. Rev.* B **81** 125120

[233] Morkoc H, Strite S, Gao G B, Lin M E, Sverdlov B and Burns M 1994 Large-band-gap SiC, III–V nitride, and II–VI ZnSe-based semiconductor-device technologies *J. Appl. Phys.* **76** 1363–98

[234] Ando T, Fowler A B and Stern F 1982 Electronic properties of two-dimensional systems *Rev. Mod. Phys.* **54** 437–72

[235] Hendry E, Hale P J, Moger J, Savchenko A K and Mikhailov S A 2010 Coherent nonlinear optical response of graphene *Phys. Rev. Lett.* **105** 097401

[236] Ferrari A C, Meyer J C, Scardaci V, Casiraghi C, Lazzeri M and Mauri F *et al* 2006 Raman spectrum of graphene and graphene layers *Phys. Rev. Lett.* **97** 187401

[237] Nyakiti L O, Myers-Ward R L, Wheeler V D, Imhoff E A, Bezares F J and Chun H *et al* 2012 Bilayer graphene grown on 4H-SiC (0001) step-free mesas *Nano Lett.* **12** 1749–56

[238] Lui C H, Li Z, Chen Z, Klimov P V, Brus L E and Heinz T F 2011 Imaging stacking order in few-layer graphene *Nano Lett.* **11** 164–9

[239] Cancado L G, Takai K, Enoki T, Endo M, Kim Y A and Mizusaki H *et al* 2008 Measuring the degree of stacking order in graphite by raman spectroscopy *Carbon* **46** 272–5

[240] Fang W, Hsu A L, Caudillo R, Song Y, Birdwell A G and Zakar E *et al* 2013 Rapid identification of stacking orientation in isotopically labeled chemical-vapor grown bilayer graphene by raman spectroscopy *Nano Lett.* **13** 1541–8

[241] Poncharal P, Ayari A, Michel T and Sauvajol J L 2008 Raman spectra of misoriented bilayer graphene *Phys. Rev.* B **78** 113407

[242] Malard L M, Nilsson J, Elias D C, Brant J C, Plentz F and Alves E S *et al* 2007 Probing the electronic structure of bilayer graphene by Raman scattering *Phys. Rev.* B **76** 201401

[243] Xie L M, Wang H L, Jin C H, Wang X R, Jiao L Y and Suenaga K *et al* 2011 Graphene nanoribbons from unzipped carbon nanotubes: atomic structures, Raman spectroscopy, and electrical properties *J. Am. Chem. Soc.* **133** 10394–7

[244] Matsui T, Kambara H, Niimi Y, Tagami K, Tsukada M and Fukuyama H 2005 STS observations of Landau levels at graphite surfaces *Phys. Rev.* B **94** 226403

[245] Uchoa B, Yang L, Tsai S W, Peres N M and Castro Neto A H 2009 Theory of scanning tunneling spectroscopy of magnetic adatoms in graphene *Phys. Rev. Lett.* **103** 206804

[246] Miller D L, Kubista K D, Rutter G M, Ruan M, de Heer W A and First P N *et al* 2009 Observing the quantization of zero mass carriers in graphene *Science* **324** 924–7

[247] Song Y J, Otte A F, Kuk Y, Hu Y, Torrance D B and First P N *et al* 2010 High-resolution tunnelling spectroscopy of a graphene quartet *Nature* **467** 185–9

[248] Jung S, Rutter G M, Klimov N N, Newell D B, Calizo I and Hight-Walker AR *et al* 2011 Evolution of microscopic localization in graphene in a magnetic field from scattering resonances to quantum dots *Nat. Phys.* **7** 245–51

[249] Chae J, Jung S, Young A F, Dean C R, Wang L and Gao Y *et al* 2012 Renormalization of the graphene dispersion velocity determined from scanning tunneling spectroscopy *Phys. Rev. Lett.* **109** 116802

[250] Li G, Luican A and Andrei E Y 2009 Scanning tunneling spectroscopy of graphene on graphite *Phys. Rev. Lett.* **102** 176804

[251] Rutter G M, Jung S, Klimov N N, Newell D B, Zhitenev N B and Stroscio J A 2011 Microscopic polarization in bilayer graphene *Nat. Phys.* **7** 649–55

[252] Miller D L, Kubista K D, Rutter G M, Ruan M, de Heer W A and Kindermann M *et al* 2010 Real-space mapping of magnetically quantized graphene states *Nat. Phys.* **6** 811–7
[253] Hashimoto K, Champel T, Florens S, Sohrmann C, Wiebe J and Hirayama Y *et al* 2012 Robust nodal structure of Landau level wave functions revealed by Fourier transform scanning tunneling spectroscopy *Phys. Rev. Lett.* **109** 116805
[254] Fu Y S, Kawamura M, Igarashi K, Takagi H, Hanaguri T and Sasagawa T 2014 Imaging the two-component nature of Dirac–Landau levels in the topological surface state of Bi_2Se_3 *Nat. Phys.* **10** 815–9
[255] Chang C P, Lu C L, Shyu F L, Chen R B, Huang Y C and Lin M F 2005 Magnetoelectronic properties of AB-stacked graphite *Carbon* **43** 1424–31
[256] Lu C L, Chang C P and Lin M F 2007 Magneto-electronic properties of AA- and ABC-stacked graphites *Eur. Phys. J.* B **60** 161–9

Chiun-Yan Lin, Thi-Nga Do, Yao-Kung Huang and Ming-Fa Lin

Chapter 2

The generalized tight-binding model

The primitive unit cell of the honeycomb structure of monolayer graphene, marked by the gray diamond, possesses two carbon atoms A (black) and B (red) atoms, shown in figure 2.1(a). The symbols γ_0 ($\simeq -2.6$ eV) and b (=1.42 Å) indicate the nearest-neighbor hopping integral and C–C bond length, respectively. The first Brillouin zone is shown in figure 2.1(b), where Γ, M and K are three highly symmetric points. The low-energy band structure is dominated by the π-bonding perpendicular to the graphene plane. In the tight-binding model based on the $2p_z$ orbitals, the Hamiltonian is a 2 × 2 Hermitian matrix. The Bloch wavefunction can be described by a linear combination of the two tight-binding functions of the A and B sublattices (φ_A and φ_B) [1]:

$$\Psi = \varphi_A + \lambda\varphi_B = \sum \exp(\mathrm{i}\mathbf{k}\cdot\mathbf{R}_A)\chi(\mathbf{r}-\mathbf{R}_A) + \lambda\exp(\mathrm{i}\mathbf{k}\cdot\mathbf{R}_B)\chi(\mathbf{r}-\mathbf{R}_B), \quad (2.1)$$

where $\lambda = \pm 1$ indicates the bonding and anti-bonding, and $\chi(\mathbf{r})$ is the normalized $2p_z$ orbital wavefunction for an isolated carbon atom. The 2 × 2 tight-binding Hamiltonian in the nearest-neighbor approximation is expressed as:

$$\begin{bmatrix} 0 & \gamma_0 f(k) \\ \gamma_0 f(k)^* & 0 \end{bmatrix}, \quad (2.2)$$

where $f(k) = \mathrm{e}^{\mathrm{i}}k_x b/\sqrt{3} + 2\mathrm{e}^{-\mathrm{i}k_x b/2\sqrt{3}}\cos(\frac{k_y b}{2})$, and the site energies of the A and B sublattices and the overlap integral between them are set to zero.

The optical response due to an electromagnetic field is described by the absorption spectral function $A(\omega)$. According to Fermi's golden rule, at zero temperature $A(\omega)$ is expressed as:

doi:10.1088/978-0-7503-1566-1ch2 © IOP Publishing Ltd 2017

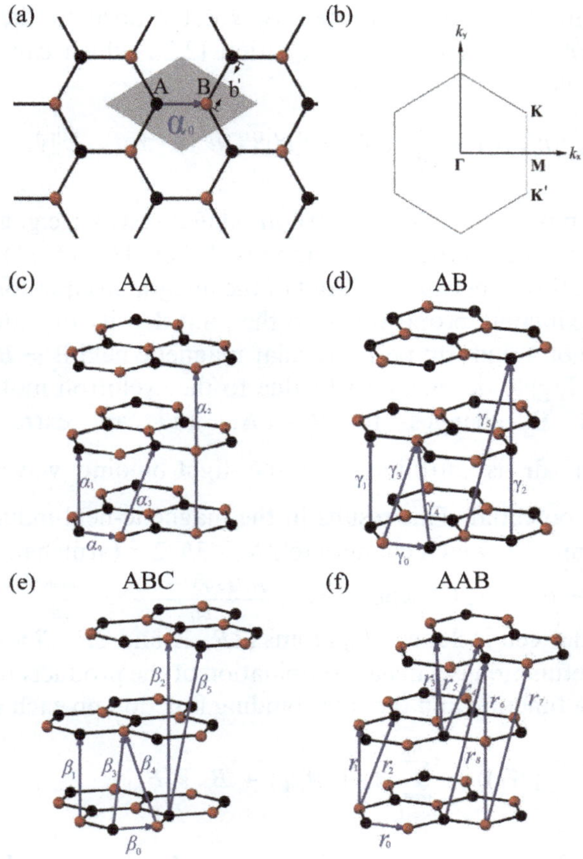

Figure 2.1. (a) The honeycomb lattice structure of monolayer graphene. Sublattices A and B are shown in black and red, respectively. The primitive unit cell is depicted by a gray diamond, where α indicates the nearest-neighbor hopping integral and b' represents the C–C bond length. (b) The Brillouin zone of the honeycomb lattice and some highly symmetric points. The geometric structures of (c) AA-, (d) AB-, (e) ABC- and (f) AAB-stacked graphene; the inter- and intralayer atomic interactions are indicated by blue arrows.

$$A(\omega) \propto \sum_{n^v, n^c} \int_{1stBZ} \frac{d\mathbf{k}}{2\pi * 2\pi} \left| \left\langle \Psi^c_{\mathbf{k}}(n^c) \middle| \frac{\hat{\mathbf{E}} \cdot \mathbf{P}}{m_e} \middle| \Psi^v_{\mathbf{k}}(n^v) \right\rangle \right|^2 \times Im \left\{ \frac{f\left[E^c_{\mathbf{k}}(n^c)\right] - f\left[E^v_{\mathbf{k}}(n^v)\right]}{E^c_{\mathbf{k}}(n^c) - E^v_{\mathbf{k}}(n^v) - \omega - \imath\Gamma} \right\}, \tag{2.3}$$

where $\mathbf{E}$ is the electric polarization vector, $\mathbf{P}$ is the momentum operator, $f[E_{k_y}(n)]$ is the Fermi–Dirac distribution, m_e is the electron mass and Γ is the phenomenological broadening parameter. Only vertical transitions ($\Delta k = 0$) are available from the initial occupied state n^v to the final unoccupied state n^c because of the vanishing photon momentum. The transition intensity is determined by the velocity matrix element $M_{\mathbf{k}}^{c,v}(n^c, n^v)$, the first term in equation (2.2), which can be evaluated as follows in the gradient approximation:

$$M_{\mathbf{k}}^{c,v}(n^c, n^v) \sim \frac{\partial}{\partial k_y}\left\langle \Psi_{\mathbf{k}}^c(n^c)\middle|H\middle|\Psi_{\mathbf{k}}^v(n^v)\right\rangle \quad \text{for} \quad \hat{E}\|\hat{y}. \tag{2.4}$$

The approximation is well verified for carbon-related systems, e.g. graphene, carbon nanotubes, graphite and nanographite ribbons. When $M_{\mathbf{k}}^{c,v}(n^c, n^v)$ slowly varies in the Brillouin zone, this term can be left out of the integral in equation (2.3), and $A(\omega)$ is therefore approximately proportional to the joint density of states.

In the presence of a uniform perpendicular magnetic field $\mathbf{B} = B_0\hat{z}$ the electronic states evolve into highly degenerate LLs due to the cyclotron motion of electrons. The momentum $\mathbf{P}$ changes to $\mathbf{P} - e\mathbf{A}$, and an extra Peierls phase $G_R \equiv \frac{2\pi}{\phi_0}\int_{\mathbf{R}'}^{\mathbf{R}} \mathbf{A}(\mathbf{r}) \cdot \mathbf{dr}$ is introduced to the tight-binding wavefunction, where $\mathbf{A}(r)$ is the vector potential. This results in the magnetic-field-induced enlargement of the primitive unit cell, which changes to $2R_B \cdot 3b'/2\ \hat{x}$ (armchair direction) in the Landau gauge $A = (0, B_0\mathrm{x}, 0)$, where $R_B = \frac{\phi_0/(3\sqrt{3}b'^2/2)}{B_0} \simeq \frac{79000\ \mathrm{T}}{B_0}$. Accordingly, the enlarged rectangular cell includes $4R_B$ atoms ($2R_B$ A and $2R_B$ B atoms). The Bloch tight-binding wavefunction is a linear combination of the products between the value of the subenvelope function and the tight-binding function on each sublattice site:

$$|\ \Psi_{\mathbf{k}})\rangle = \sum_{m=1}^{2R_B} A_m\ |\ A_{m\mathbf{k}}\rangle + B_m\ |\ B_{m\mathbf{k}}\rangle, \tag{2.5}$$

where A_m and B_m are the subenvelope functions and m indicates the atom number. The quantum number of each Landau state can be characterized by the numbers of the zero points of the dominating subenvelope wavefunction. The Hamiltonian can be expressed as a band-symmetric matrix in the bases of $\{|A_{1\mathbf{k}}\rangle, |B_{2R\mathbf{k}}\rangle, |B_{1\mathbf{k}}\rangle, |A_{2R\mathbf{k}}\rangle, |A_{2\mathbf{k}}\rangle, |B_{2R-1\mathbf{k}}\rangle, |B_{2\mathbf{k}}\rangle, |A_{2R-1\mathbf{k}}\rangle, \ldots |A_{R-1\mathbf{k}}\rangle, |B_{R+2\mathbf{k}}\rangle, |B_{R-1\mathbf{k}}\rangle, |A_{R+2\mathbf{k}}\rangle, |A_{R\mathbf{k}}\rangle, |B_{R+1\mathbf{k}}\rangle, |B_{R\mathbf{k}}\rangle, |A_{R+1\mathbf{k}}\rangle\}$ that is,

$$\begin{bmatrix} 0 & q^* & p_1^* & 0 & \cdots & 0 & 0 & 0 \\ q & 0 & 0 & p_{2R} & 0 & \cdots & 0 & 0 \\ p_1 & 0 & 0 & 0 & \ddots & \ddots & \vdots & 0 \\ 0 & p_{2R}^* & 0 & 0 & \ddots & \ddots & 0 & \vdots \\ \vdots & 0 & \ddots & \ddots & \ddots & 0 & p_R^* & 0 \\ 0 & \vdots & \ddots & \ddots & 0 & 0 & 0 & p_{R+1} \\ 0 & 0 & \cdots & 0 & p_R & 0 & 0 & q \\ 0 & 0 & 0 & \cdots & 0 & p_{R+1}^* & q^* & 0 \end{bmatrix}. \tag{2.6}$$

The independent elements p_m and q are defined as

$$\begin{aligned} p_m &\equiv \gamma_0 \exp(-ik_x b/2) \cdot 2\cos(k_y\sqrt{3}b/2 + \Delta G_{mm'}), \\ q &\equiv \gamma_0 \exp(ik_x b); \\ \Delta G_{mm'} &= -\pi(m - 5/6)/R. \end{aligned} \tag{2.7}$$

Substituting equation (2.5) and the derivative of equation (2.6) into equation (2.4), one obtains the velocity matrix $M_{\mathbf{k}}^{c,v}(n^c, n^v) =$

$$\begin{aligned} &\left[A_1^{c*}\; B_{2R}^{c}{}^*\; B_1^{c*} \cdots B_{R+1}^{c}{}^*\; B_R^{c*}\; A_{R+1}^{c}{}^*\right] \begin{bmatrix} 0 & 0 & u_1^* & 0 & \cdots & 0 & 0 & 0 \\ 0 & 0 & 0 & u_{2R} & 0 & \cdots & 0 & 0 \\ u_1 & 0 & 0 & 0 & \ddots & \ddots & \vdots & 0 \\ 0 & u_{2R}^* & 0 & 0 & \ddots & \ddots & 0 & \vdots \\ \vdots & 0 & \ddots & \ddots & \ddots & 0 & u_R^* & 0 \\ 0 & \vdots & \ddots & \ddots & 0 & 0 & 0 & u_{R+1} \\ 0 & 0 & \cdots & 0 & u_R & 0 & 0 & v \\ 0 & 0 & 0 & \cdots & 0 & u_{R+1}^* & 0 & 0 \end{bmatrix} \begin{bmatrix} A_1^v \\ B_{2R}^v \\ B_1^v \\ \vdots \\ \vdots \\ B_{R+1}^v \\ B_R^v \\ A_{R+1}^v \end{bmatrix} \\ &= A_1^{c*}B_1^v u_1^* + A_2^{c*}B_2^v u_2^* + \cdots + A_{2R}^{c}{}^* B_{2R}^v u_{2R}^* + B_1^{c*}A_1^v u_1 + B_2^{c*}A_2^v u_2 + \cdots \\ &\quad + B_{2R}^{c}{}^* A_{2R}^v u_{2R} \\ &= \sum_{m=1}^{2R_B} A_m^{c*}B_m^v u_m^* + B_m^{c*}A_m^v u_m, \end{aligned} \tag{2.8}$$

where the element u_m is given by

$$u_m \equiv -\sqrt{3}b\gamma_0 \exp(-ik_x b/2) \cdot 2\sin(k_y\sqrt{3}b/2 + \Delta G_{mm'}). \tag{2.9}$$

Considering the transition intensity of the optical excitations in equation (2.4), one can move the Peierls phase term out of the summation because it changes slowly in the enlarged unit cell. As a consequence, the transition intensity $M_{\mathbf{k}}^{c,v}(n^c, n^v)$ is determined by the inner product of the different subenvelope functions of the initial and final states, i.e. it is proportional to $c_1A_m^cB_m^v + c_2B_m^cA_m^v$, where c_1 and c_2 are constants. According to the orthogonal relationship of the subenvelope functions, the selection rule is deduced to be $\Delta n = \pm 1$.

The generalized tight-binding model, based on the subenvelope functions of the distinct sublattices, is suitable for studying the magneto-electronic properties of multilayer graphene with arbitrary stacking. The effects of external fields and atomic interactions are simultaneously taken into account. When an electric field is perpendicularly applied to the graphene planes, the electric potentials are added to the site energies of the diagonal elements of the Hamiltonian matrix. This model provides an exact diagonalization method for a band-like Hamiltonian matrix to efficiently solve the eigenvalues and eigenfunctions of the quantized states. It can also be utilized to comprehend the essential properties of the non-uniform or composite external fields [2]. As for the absorption spectra, they are closely related to the Landau-level spectrum and the transition matrix elements. The spectral characteristics, such as the prominent structures, number, intensity and frequency of the absorption peaks, are well depicted. Moreover, the results are accurate and reliable within a wide frequency range.

The velocity matrix, which determines the transition intensity and the selection rule, deserves a closer examination. It should also be noted that the velocity matrix needs to be calculated from the wavefunctions of the initial and final states, of which the individual wavefunction is a linear combination of the products between the subenvelope function and the tight-binding function on each site in the magnetically enlarged unit cell. Using the gradient approximation, one can obtain that the expectation value of the velocity matrix is simplified as the product of three matrices: the subenvelope functions of the initial state and the final state, and the operator $\frac{\partial H}{\partial k}$. The calculation result is proportional to the sum of the products of the components of the the initial and final states correlating to the non-vanishing hopping integrals. It can be deduced that for multilayer graphene, the terms attributed to the in-plane atomic interaction γ_0 dominate the optical excitations because intralayer atomic interactions are much stronger than interlayer ones. Consequently, the available optical transitions satisfy the condition in which the quantum mode needs to be identical for the A (B) sublattice of the initial state and the B (A) sublattice of the final state on the same layer, according to the orthogonal relationship of the subenvelope functions. Detailed calculations are given in the following sections, as well as clear physical pictures for explaining the magneto-optical properties.

The generalized tight-binding model, which is based on the subenvelope functions of distinct sublattices, is suitable for exploring the diverse quantization phenomena arising from the various stacking configurations, layer numbers, dimensions and external fields, e.g. the rich magnetic quantization of the multi-constant-energy surface in ABC, AAB and sliding stacking. Moreover, this model can combine with the many-particle theories to study other essential properties, e.g. Coulomb excitations [3–9]. On the other hand, most research groups use the effective mass model to deal with the electronic properties in external fields. These two models are quite different from each other. For example, the dimension of the magnetic Hamiltonian is huge for the generalized tight-binding model, while it is very small and the same as that of the zero-field for the effective mass model. However, the high barriers were created by this low-energy expansion method; that is, it is very difficult to explore the rich and complex magnetic quantization and the interlayer atomic interactions in typical stacking systems using this approximation, as discussed in the previously published papers [10–13].

In the following chapters, five kinds of typical stacking—AAA, ABA, ABC, AAB and sliding—are chosen for systematic study, since they can exhibit diverse phenomena under external fields. Physical pictures relating to the stacking configurations are proposed to characterize the generalized tight-binding model, and the final formula is expressed in equations. These layered graphene systems should be suitable for a detailed comparison with the graphene-related systems. The important dimension-induced differences between 3D graphite, 2D graphene, 1D graphene nanoribbons and 1D carbon nanotubes are discussed in chapter 9. This should be helpful for readers, and the differences can be further generalized to other layered systems, e.g. MoS_2-related 2D systems and groups IV and V layered systems.

References

[1] Wallace P R 1947 The band theory of graphite *Phys. Rev.* **71** 622–34

[2] Ou Y C, Chiu Y H, Yang P H and Lin M F 2014 The selection rule of graphene in a composite magnetic field *Opt. Exp.* **22** 7473–91

[3] Ho J H, Lu C L, Hwang C C, Chang C P and Lin M F 2006 Coulomb excitations in AA- and AB-stacked bilayer graphites *Phys. Rev.* B **74** 085406

[4] Ho J H, Chang C P and Lin M F 2005 Electronic excitations of the multilayered graphite *Phys. Lett.* A **352** 446–50

[5] Lin M F, Chuang Y C and Wu J Y 2012 Electrically tunable plasma excitations in AA-stacked multilayer graphene *Phys. Rev.* B **86** 125434

[6] Wu J Y, Chen S C, Roslyak O, Gumbs G and Lin M F 2011 Plasma excitations in graphene: their spectral intensity and temperature dependence in magnetic field *ACS Nano* **5** 1026–32

[7] Lozovik Y E and Sokolik A A 2012 Influence of Landau level mixing on the properties of elementary excitations in graphene in strong magnetic field *Nano. Res. Lett.* **7** 134

[8] Wu J Y, Gumbs G and Lin M F 2014 Combined effect of stacking and magnetic field on plasmon excitations in bilayer graphene *Phys. Rev.* B **89** 165407

[9] Wu J Y, Lin C Y, Gumbs G and Lin M F 2015 The effect of perpendicular electric field on temperature-induced plasmon excitations for intrinsic silicene *RSC Adv.* **5** 51912–8

[10] Do T N, Lin C Y, Lin Y P, Shih P H and Lin M F 2015 Configuration-enriched magneto-electronic spectra of AAB-stacked trilayer graphene *Carbon* **94** 619–32

[11] Huang Y K, Chen S C, Ho Y H, Lin C Y and Lin M F 2014 Feature-rich magnetic quantization in sliding bilayer graphenes *Sci. Rep.* **4** 7509

[12] Lin C Y, Wu J Y, Ou Y J, Chiu Y H and Lin M F 2015 Magneto-electronic properties of multilayer graphenes *Phys. Chem. Chem. Phys.* **17** 26008–35

[13] Lin C Y, Wu J Y, Chiu Y H, Chang C P and Lin M F 2014 Stacking-dependent magnetoelectronic properties in multilayer graphene *Phys. Rev.* B **90** 205434

Chiun-Yan Lin, Thi-Nga Do, Yao-Kung Huang and Ming-Fa Lin

Chapter 3

AA-stacked graphene

AA-stacked graphene has the highest symmetry, with the carbon atoms of one layer directly above/below those of another one. It has been artificially synthesized and is macroscopically stable, although an AA stacking configuration is absent in natural graphite due to its higher total energy [1]. Low-lying multiple Dirac cones have attracted the intense interest of theoretical and experimental research [2]. In this chapter, we focus on the optical absorption spectra in a wide energy range, and the results are closely related to the electronic band structure and the dipole transition matrix. In the absence of external fields, the low- and middle-frequency optical properties are, respectively, dominated by the excitations of massless Dirac fermions and saddle-point states. On the other hand, the magneto-optical properties are dominated by inter-LL transitions that obey a special selection rule derived from the dipole transition moment between the initial and final states. For all excitation channels, the linear symmetric and asymmetric superpositions of the wavefunctions in different layers are responsible for the non-zero and zero probability, respectively, for intra- and inter-Dirac-cone transitions, regardless of the external fields. In particular, intensity discontinuities and forbidden transition regions are revealed for certain kinds of excitation channels. The optical spectra can be further modified by the application of an electric field.

3.1 Electronic and optical properties without external fields

3.1.1 Monolayer

The electronic structure of monolayer graphene is symmetric about $E_F = 0$, as shown by the gray curve in figure 3.1(a). In the low-energy region, two Dirac-type linear subbands intersect at $E_F = 0$, with the Fermi-momentum state located at the K point. For middle energies, the saddle-point states are located at the M point with the energies of $\pm\alpha_0$. The isotropic conical dispersion measured from the K point is described by $E^{c,v} \simeq 3\alpha_0 b(\mathbf{k} - \mathbf{K})/2$ according to equation (2), where the the conical

doi:10.1088/978-0-7503-1566-1ch3 © IOP Publishing Ltd 2017

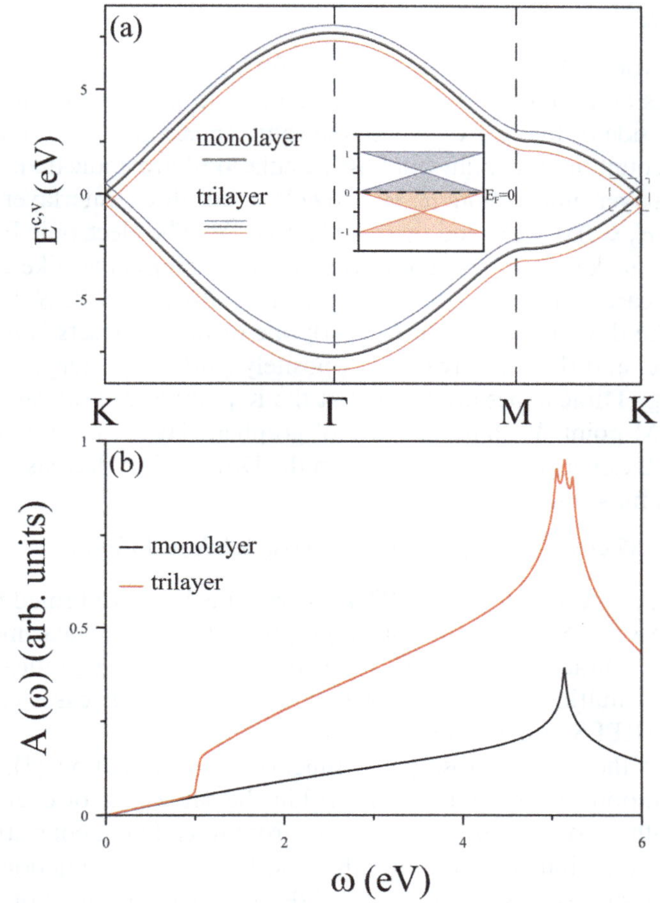

Figure 3.1. (a) The zero-field band structures and (b) the absorption spectra of monolayer and AA-stacked trilayer graphene. The shadows inset in (a) indicate the forbidden optical transition zone in the case of trilayer graphene.

point is defined as the Dirac point and the slope $\mathrm{d}E^{c,v}/\mathrm{d}k = 3\alpha_0 b/2$ is regarded as the effective speed of light of the Dirac fermions in graphene.

The main characteristics of the electronic properties are reflected in the absorption spectra. The spectral absorption function $A(\omega)$, equation (3), is obtained by a summation over all possible transitions in the first Brillouin zone. For the excitations of the Dirac fermions, $A(\omega)$ is simplified as a product of the joint density of states and the Fermi velocity. Therefore, a linear dependence of the spectral intensity on the frequency appears in the low-frequency region, while in the middle-frequency region, a symmetric peak in the logarithmically divergent form is revealed at $\omega \simeq 2\alpha_0$ due to the saddle-point transitions around the M point, as shown by the black curve in figure 3.1(b). This very strong π-electronic absorption peak is one of the critical characteristics of carbon-related sp^2 bonding systems, such as graphite [3], graphite intercalation compounds [4], carbon nanotubes [5, 6] and carbon fullerenes [7].

3.1.2 Multilayer

Highly symmetric AA-stacked graphene is formed by perpendicularly stacking graphene sheets along the $\hat{z}$ direction. The geometric structure is illustrated in the top right-hand side of figure 1(c), where the four representative atomic interactions taken into account in the tight-binding calculations are indicated: one nearest-neighbor intralayer interaction $\alpha_0 = 2.569$ eV, and three interlayer interactions $\alpha_1 = 0.361$ eV, $\alpha_2 = 0.013$ eV and $\alpha_3 = -0.032$ eV [8]. The electronic band structure of N-layer AA stacked graphene comprises N sets of monolayer-like energy bands. In the trilayer case, three sets of monolayer-like subbands, $S_1^{c,v}$, $S_2^{c,v}$ and $S_3^{c,v}$, are displayed by the thin curves in figure 3.1(a); one of the three sets is mapped to the monolayer case, and the other two are oppositely shifted by energy $\sim\alpha_1$. Note that three low-energy Dirac cones are located at the K point and three saddle points are located at the M point. With an increase of graphene layers, more Dirac cones are induced, while the energy difference between the Dirac points decreases, as described in the effective mass model by [9]

$$[2\cos(j + 1\pi/(N + 1)) - 2\cos(j\pi/(N + 1))]\alpha_1, \tag{3.1}$$

where $j = 1, 2, \ldots, N - 1$. Recently, STS experiments have confirmed that there are two Dirac cones in AA-stacked bilayer graphene [10]. The measured slope and location of the Dirac cones are in agreement with the theoretical predictions [2, 11, 12]. The multiple Dirac cone structure for multilayer cases can be further confirmed by ARPES and STS experiments.

According to the spectral absorption function $A(\omega)$, equation (3), we find that interband excitations are only available within the same pair of energy subbands. That is to say, they come from intra- but not from inter-Dirac-cone structures. The reason for this is attributed to the fact that the Landau wavefunction is the linear symmetric or asymmetric superposition of the subenvelope functions in different layers, which leads to non-zero (zero) electric dipole moments for intra-Dirac-cone (inter-Dirac-cone) transitions, regardless of the external perpendicular electric and magnetic fields. However, a forbidden transition zone, as indicated by shadows for the case of trilayer graphene in figure 3.1(a), is expected for the lower and upper Dirac cones because of their respective free electrons and holes according to the Fermi–Dirac distribution. This would lead to a discontinuity in the optical spectrum.

As a result of Dirac fermion excitations, AA-stacked trilayer graphene also shows a linearly dependent absorption intensity in the low-frequency spectrum (red curve in figure 3.1(b)). However, it should be noticed that due to the forbidden transition zones around the Fermi energy (the shadows in figure 3.1(a)), a shoulder structure occurs near $\omega \simeq 1$ eV, which is evaluated as double the energy difference between $E_F = 0$ and the lower (higher) Dirac point energy or the difference between the two Dirac point energies. The measured forbidden zone can be used to estimate the value of α_1 according to equation (3.1). The spectral intensity for $\omega < 1$ eV is equal to the intensity of monolayer graphene, because only the middle Dirac cone contributes to the optical transitions. On the other hand, for $\omega > 1$ eV, interband transitions within all three Dirac cones are permitted. Therefore, the linear intensity is three times as large as that of monolayer graphene. In the middle-frequency region $\sim 2\alpha_0$, instead of one single peak for monolayer graphene, three close peaks are induced as a result of more excitation channels around the saddle-points at the M point. An increase of graphene layers gives rise to verified and diversified spectral structures in which more steps and peaks appear at the low and middle frequencies respectively. It should be noted that even-layer AA-stacked graphene possesses a bilayer-like optical gap determined by the spacing between the valence and conduction Dirac points nearest to the Fermi level. Previous works [12] have shown the layer-dependent spectral structures for odd-layer and even-layer scenarios. Nevertheless, for certain frequency regions, the cases in both scenarios exhibit properties identical to those of 3D graphite as the layer number exceeds the critical values [12].

3.2 Magneto-electronic and magneto-optical properties

3.2.1 Monolayer

In the presence of a magnetic field, the electronic states congregate to form dispersionless LLs. Each LL in monolayer graphene is four-fold degenerate without the consideration of spin degeneracy. For the states at ($k_x = 0$, $k_y = 0$), their corresponding wavefunctions are localized around four different centers: the 1/6, 2/6, 4/6 and 5/6 positions of the enlarged unit cell. At $B_0 = 40$ T, the LLs and the localized wavefunctions are shown in figure 3.2(a) and (b), where only the 1/6-localized Landau states are discussed. In the LL energy spectrum in figure 3.2(c), the conduction and valence LLs are symmetric about $E_F = 0$, and characterized by the quantum numbers $n^{c,v}$ corresponding to the numbers of the zero points of the subenvelope functions ($A_o^{c,v}$ for 1/6-localized states). The $n^{c,v} = 0$ LL (at $E_F = 0$) is constant as a function of B_0, and the $n^{c,v} = 1, 2, 3, \ldots$ LLs are counted away from $E_F = 0$; they follow the simple square root relationship $|E_n^{c,v}| \propto \sqrt{n^{c,v} B_0}$, which is responsible for Dirac-type linear dispersion.

The Landau wavefunctions, characterized by the subenvelope functions on different sublattices, are used to define the quantum numbers of LLs. The wavefunction can be decomposed into odd-indexed and even-indexed components in the magnetically enlarged unit cell as follows:

$$|\Psi_{\mathbf{k}}\rangle = \sum_{m=1}^{2R_B-1} (A_o|A_{m\mathbf{k}}\rangle + B_o|B_{m\mathbf{k}}\rangle) + \sum_{m=1}^{2R_B} (A_e|A_{m\mathbf{k}}\rangle + B_e|B_{m\mathbf{k}}\rangle), \tag{3.2}$$

where $o(e)$ represents an odd (even) integer. The subenvelope functions $A_{o,e}$ ($B_{o,e}$), standing for the probability amplitudes, exhibit even or odd spatial symmetry and can be described by an nth order Hermite polynomial multiplied by a Gaussian function. The amplitudes are identical for the 1/6 and 4/6 localized wavefunctions, and so are for those for the 2/6 and 5/6 functions. The $A_{o,e}^{c,v}$ of the 1/6 localized state is equal to the $B_{o,e}^{c,v}$ of the 2/6 localized state and vice versa. Moreover, the odd-indexed component only has a phase difference of π with respect to the even-indexed component, i.e. $A_o^{c,v} = -A_e^{c,v}$ and $B_o^{c,v} = -B_e^{c,v}$. Therefore, discussing the

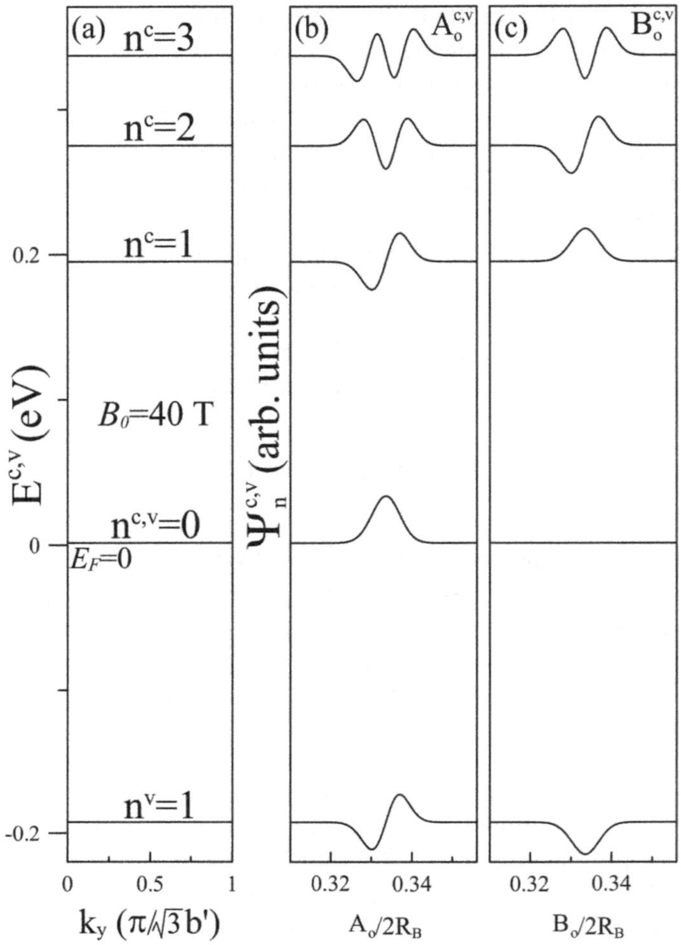

Figure 3.2. Low-lying Landau levels of monolayer graphene at $B_0 = 40$ T. The quantum numbers of the conduction and valence Landau levels $n_{c,v} = 0, 1, 2, 3...$ are counted from the Fermi level, $E_F = 0$. (b and c) The subenvelope functions of two sublattices A and B are plotted in the unit $2R_B$ according to the Landau levels in (a).

odd-indexed components of the 2/6 localized wavefunctions is sufficient in order to comprehend the LL properties. For a Landau state with the quantum number $n^{c,v} \geqslant 1$, the subenvelope functions for the A_o and B_o sublattices correspond to the $n^{c,v}$ th and $(n^{c,v} + 1)$ th order Hermite polynomials, respectively. The conduction and valence subenvelope functions are related to each other by the following relationship: $A_o^c = A_o^v$ and $B_o^c = -B_o^v$. It can be deduced that a simple linear relationship exists between the subenvelope functions A_o and B_o of the conduction and valence Landau states, i.e. $A_o^v(n^v) \propto B_o^c(n^c = n^v - 1)$ and $B_o^v(n^v) \propto A_o^c(n^c = n^v + 1)$. This accounts for the specific selection rule $\Delta n = n^c - n^v = \pm 1$ for the inter-LL transitions, because according to equations (8) and (9), the expectation value of the velocity matrix is non-zero only when $A_o{}^v$ (B_o^v) of the initial state and $B_o{}^c$ $(A_o{}^c)$ of the final state have the same mode.

Feature-rich magneto-absorption spectra directly reflect the LL characteristics. The inter-LL transitions that obey the selection rule $\Delta n = n^c - n^v = \pm 1$ contribute to many symmetric delta-function-like absorption peaks, as shown in figure 3.3(a). For a peak denoted as $\omega_{\underline{n}(n+1)}$, the first number with a line underneath and the second number without, respectively, indicate the quantum number of the unoccupied valence LL and that of the occupied conduction LL. The absorption intensity is the same for each peak because each LL has the same degree of degeneracy. Due to the band symmetry about $E_F = 0$, the peak indicated by $\underline{n}(n+1)$ ($n = 0, 1, 2, 3, \ldots$) results from two cases of transitions: one from $n^c - n^v = 1$ with $n^v = n$ and the other from $n^c - n^v = -1$ with $n^v = n + 1$. Furthermore, the absorption frequency is

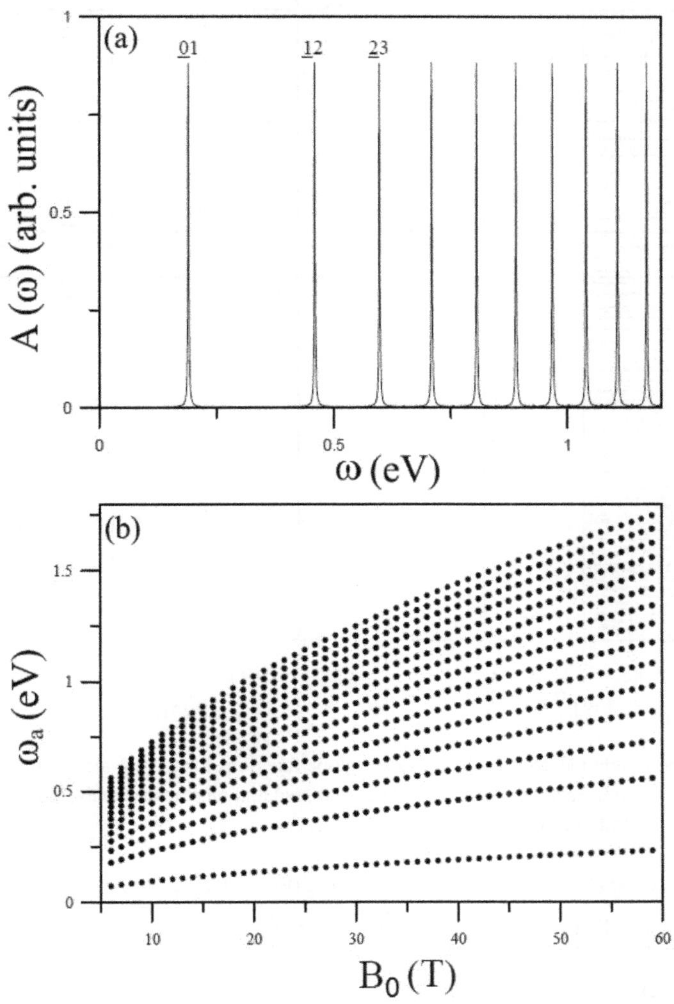

Figure 3.3. (a) The magneto-absorption spectrum of monolayer graphene at $B_0 = 40$ T. (b) The $\sqrt{B_0}$-dependent magneto-absorption frequencies.

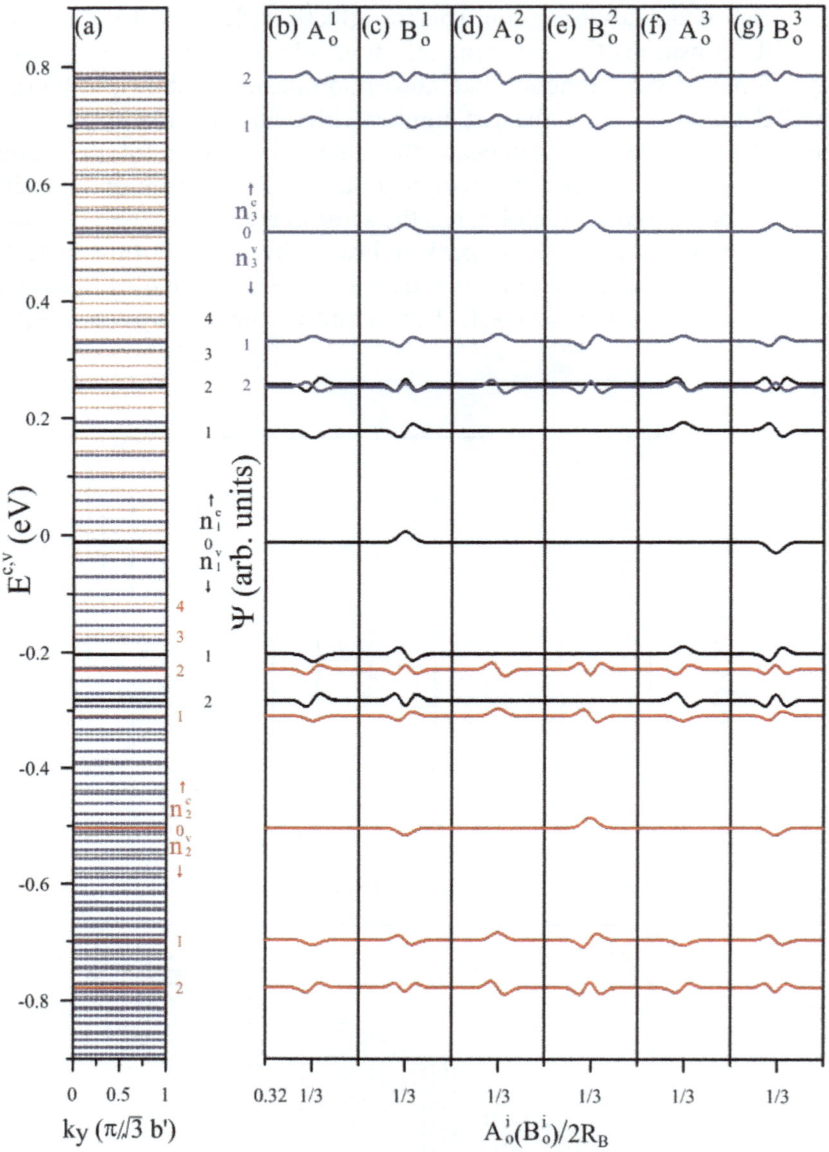

Figure 3.4. (a) The Landau levels and (b) six subenvelope functions for the 2/6 localization Landau states of AA-stacked trilayer graphene under $B_0 = 40$ T. The black, red and blue, respectively, indicate the first, second and third groups of Landau levels.

proportional to $\sqrt{B_0}$, and at a fixed magnetic field, the peaks are sequenced according to a proportional relationship $E^c - E^v \propto (\sqrt{n+1} + \sqrt{n})$ (figure 3.4(b)).

3.2.2 Multilayer

For N-layer AA-stacked graphene, the primitive unit cell under a uniform magnetic field consists of $N \times 4R_B$ atoms. In the tight-binding basis of $|A_{1\mathbf{k}}^1\rangle, |B_{1\mathbf{k}}^1\rangle, |A_{1\mathbf{k}}^2\rangle, |B_{1\mathbf{k}}^2\rangle, \ldots$

$|A_{1\mathbf{k}}^{N-1}\rangle, |B_{1\mathbf{k}}^{N-1}\rangle, |A_{1\mathbf{k}}^{N}\rangle, |B_{1\mathbf{k}}^{N}\rangle, \ldots |A_{2\mathbf{k}}^{1}\rangle, |B_{2\mathbf{k}}^{1}\rangle, |A_{2\mathbf{k}}^{2}\rangle, |B_{2\mathbf{k}}^{2}\rangle, \ldots |A_{2\mathbf{k}}^{N-1}\rangle, |B_{2\mathbf{k}}^{N-1}\rangle, |A_{2\mathbf{k}}^{N}\rangle, |B_{2\mathbf{k}}^{N}\rangle,$ $\ldots |A_{2R_B\mathbf{k}}^{1}\rangle, |B_{2R_B\mathbf{k}}^{1}\rangle, |A_{2R_B\mathbf{k}}^{2}\rangle, |B_{2R_B\mathbf{k}}^{2}\rangle, \ldots |A_{2R_B\mathbf{k}}^{N-1}\rangle, |B_{2R_B\mathbf{k}}^{N-1}\rangle, |A_{2R_B\mathbf{k}}^{N}\rangle; |B_{2R_B\mathbf{k}}^{N}\rangle$, the Peierls Hamiltonian is a band-like matrix and can be expressed as

$$\begin{aligned}\langle B_{m\mathbf{k}}^{L}|H|A_{m'\mathbf{k}}^{L'}\rangle &= \alpha_0(t_{1k}\delta_{m',m} + t_{4k}\delta_{m',m+1})\delta_{L,L'} + \alpha_3(t_{1k}\delta_{m',m} + t_{4k}\delta_{m',m-1})\delta_{L-1,L'}\\ &\quad + \alpha_3(t_{1k}\delta_{m',m} + t_{4k}\delta_{m',m+1})\delta_{L+1,L'};\\ \langle A_{m\mathbf{k}}^{L}|H|A_{m'\mathbf{k}}^{L'}\rangle &= \langle B_{m\mathbf{k}}^{L}|H|B_{m'\mathbf{k}}^{L'}\rangle = \alpha_1\delta_{L,L'+1} + \alpha_2\delta_{L,L'+2},\end{aligned} \tag{3.3}$$

where the superscript represents the layer number. The four independent phase terms are

$$\begin{aligned}t_{1k} &= \exp\left\{\mathrm{i}\left[-(k_x b/2) - (\sqrt{3}k_y b/2)\right] + \pi(m - 5/6)/R\right\}\\ &\quad + \exp\left\{\mathrm{i}\left[-(k_x b/2) + (\sqrt{3}k_y b/2)\right] - \pi(m - 5/6)/R\right\},\\ t_{2k} &= \exp\left\{\mathrm{i}\left[-(k_x b/2) - (\sqrt{3}k_y b/2)\right] + \pi(m - 3/6)/R\right\}\\ &\quad + \exp\left\{\mathrm{i}\left[-(k_x b/2) + (\sqrt{3}k_y b/2)\right] - \pi(m - 3/6)/R\right\},\\ t_{3k} &= \exp\left\{\mathrm{i}\left[-(k_x b/2) - (\sqrt{3}k_y b/2)\right] + \pi(m - 1/6)/R\right\}\\ &\quad + \exp\left\{\mathrm{i}\left[-(k_x b/2) + (\sqrt{3}k_y b/2)\right] - \pi(m - 1/6)/R\right\},\\ t_{4k} &= \exp(\mathrm{i}k_x b).\end{aligned} \tag{3.4}$$

The magnetic quantization of the N pairs of Dirac cones in N-layer AA-stacked graphene gives rise to N low-lying LL groups, of which the group divisions are based on the energy spectrum and the characteristics of the spatial wavefunction. In the case of AA-stacked trilayer graphene, the three Dirac cones at $B_0 = 40$ T evolve into three monolayer-like LL patterns, as indicated by the black, red and blue curves in figure 3.4. Each Landau state at a given (k_x, k_y) is four-fold degenerate. The conduction LLs ($E^{c}_{n_1^c=0,1,2,\ldots}$, $E^{c}_{n_2^c=0,1,2,\ldots}$, and $E^{c}_{n_3^c=0,1,2,\ldots}$) are approximately symmetric to the valence LLs ($E^{v}_{n_1^v=0,1,2,\ldots}$, $E^{v}_{n_2^v=0,1,2,\ldots}$ and $E^{v}_{n_3^v=0,1,2,\ldots}$) with respect to the zero-mode LL, as a result of the three isotropic Dirac cones. The zero-mode energies $E^{c,v}_{n_1^{c,v}=0}$, $E^{c,v}_{n_2^{c,v}=0}$ and $E^{c,v}_{n_3^{c,v}=0}$ are revealed to be the energies of the three Dirac points. The two subenvelope functions in the first group have the same relationship with monolayer graphene for each layer, and so does the third group. It should be noted that the second group of LLs has the same relationship only in the two outermost layers, while the innerlayer subenvelope functions are empty. This reflects the fact that the second group of LLs is regarded as a result of monolayer graphene, which is consistently presented when the number of graphene layers is odd. This is a visible means of identifying the LLs in terms of single-mode wavefunctions, since there is an absence of LL couplings for the AA stacking configuration.

The energy distributions of three groups of LLs are related to the individual Dirac cones that are primarily separated by the specific interlayer atomic interaction α_1 [2, 9, 13]. Each group exhibits a simple square root dependence on the field strength and the quantum number, like monolayer graphene. The energies of the zero-mode LLs are

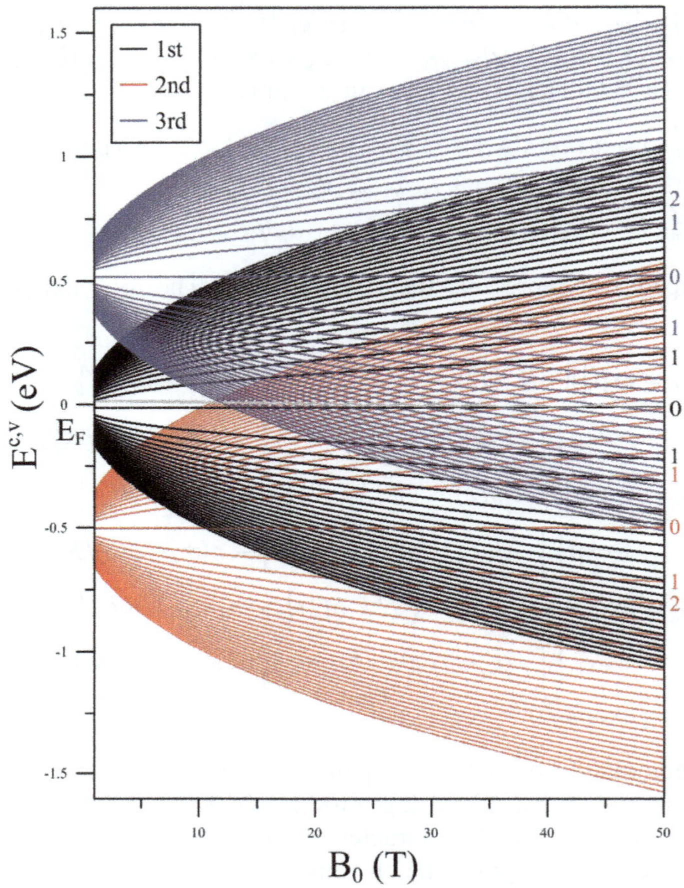

Figure 3.5. The B_0-dependent LL energy spectrum of AA-stacked trilayer graphene. The three monolayer-like LL patterns, as indicated in black, red and blue, evolve from the three isolated Dirac cones in figure 3.1(a).

constant as a function of the field strength; each LL moves towards the individual Dirac point energy when the magnetic field approaches zero. The Fermi energy for a constant carrier density oscillates as a function of the magnetic field strength around the zero-field value of $E_F = 0$ (the bold wriggling curves around $E = 0$ in figure 3.5), leading to an oscillating energy gap and a metal–semiconductor transition. The magneto-resistance measurements are used to detect the oscillation of the Fermi level for other 2D systems, such as the FeSe and AlGaAs/GaAs heterojunctions [14–16]. The Fermi distribution changes with the oscillating Fermi energy, which might cause the change of the threshold channel and frequency for each category of intragroup LL transitions and induce a forbidden region for the optical transitions. Meanwhile, all the LLs directly cross one another—i.e. in the absence of LL anticrossing—and no additional selection rule is induced, since each Landau state is characterized by a single-mode harmonic function.

In the presence of a magnetic field, the velocity matrix elements are expressed as follows in the gradient approximation:

$$
\begin{aligned}
M_{\mathbf{k}}^{c,v}(n^c, n^v) &\cong \frac{\partial}{\partial k_y}\langle \Psi_{\mathbf{k}}^c(n^c)|H|\Psi_{\mathbf{k}}^v(n^v)\rangle \\
&= \sum_{l,l'=1}^{3}\sum_{m,m'=1}^{2R_B}\Big(c^*_{A^l_{m,k}} c_{A^{l'}_{m',k}} \frac{\partial}{\partial k_y}\langle A^l_{m,k}|H|A^{l'}_{m',k}\rangle \\
&+ c^*_{A^l_{m,k}} c_{B^{l'}_{m',k}} \frac{\partial}{\partial k_y}\langle A^l_{m,k}|H|B^{l'}_{m',k}\rangle \\
&+ c^*_{B^l_{m,k}} c_{A^{l'}_{m',k}} \frac{\partial}{\partial k_y}\langle B^l_{m,k}|H|A^{l'}_{m',k}\rangle \\
&+ c^*_{B^l_{m,k}} c_{B^{l'}_{m',k}} \frac{\partial}{\partial k_y}\langle B^l_{m,k}|H|B^{l'}_{m',k}\rangle\Big),
\end{aligned} \tag{3.5}
$$

where they are evaluated from the derivative of equation (12), and the initial- and final-state wavefunctions. Substituting the velocity matrix into the spectral absorption function $A(\omega)$, equation (3), one finds that the terms correlated by the in-plane atomic interaction play an important role, and the optical LL transitions are available only within the same group (those between the intergroup LLs are forbidden) [17]. Given that each LL behaves like the massless Dirac LL, the selection rules of the intragroup-LL transitions can be ascribed to $\Delta n = \pm 1$. On the other hand, the prohibition of intergroup-LL transitions can be attributed to the fact that the Landau wavefunctions are the linear symmetric or asymmetric superpositions of the subenvelope functions in different layers, giving rise to a zero electric dipole moment for the intergroup transitions.

The prominent delta-function-like absorption peaks can be classified into N categories, each of which is owed to the individual intragroup-LL transitions and satisfies the particular selection rule $\Delta n = \pm 1$ (the same as the monolayer). In the case of $N = 3$, the first, second and third categories of the magneto absorption peaks are indicated with black, red and blue numbers, respectively, as shown in figure 3.6(a) for $B_0 = 40$ T. The distributions of the peak frequency and intensity with respect to each category are based on the Fermi level and the Fermi–Dirac distributions of the conduction and valence LLs. As a result, the second category corresponds to a monolayer-like spectrum. Nevertheless, the first and third categories can be regarded as a bilayer-like spectrum [19, 20]. Outside the threshold peaks, $\omega_1^{8^v7^v}$ and $\omega_3^{6^c7^c}$, there exists a forbidden transition region within these two categories as a result of free carriers in the Dirac cones. The forbidden region, enclosed by two peaks excited from the Fermi level, $\omega_1^{8^v7^v}$ and $\omega_1^{8^v7^c}$; $\omega_3^{6^c7^c}$, and $\omega_3^{6^v7^c}$, is evaluated as a double of the excitation frequency between the highest occupied LL and the Fermi level, either for the first or third category. In general, two kinds of channels can contribute to one excitation peak due to the symmetric distribution of the conduction and valence LLs, e.g. $n_i^c = n \rightarrow n_i^v = n + 1$ and $n_i^v = n + 1 \rightarrow n_i^v = n$ for $\omega_i^{n,n+1}$, where index n equals 0, 1, 2, ... and i represents a group number. The same kind of permitted channels from three intragroup categories, i.e. $\omega_1^{n,n+1}$, $\omega_2^{n,n+1}$ and $\omega_3^{n,n+1}$, results in three close peaks, between which the slightly different frequency reflects the distortion of the

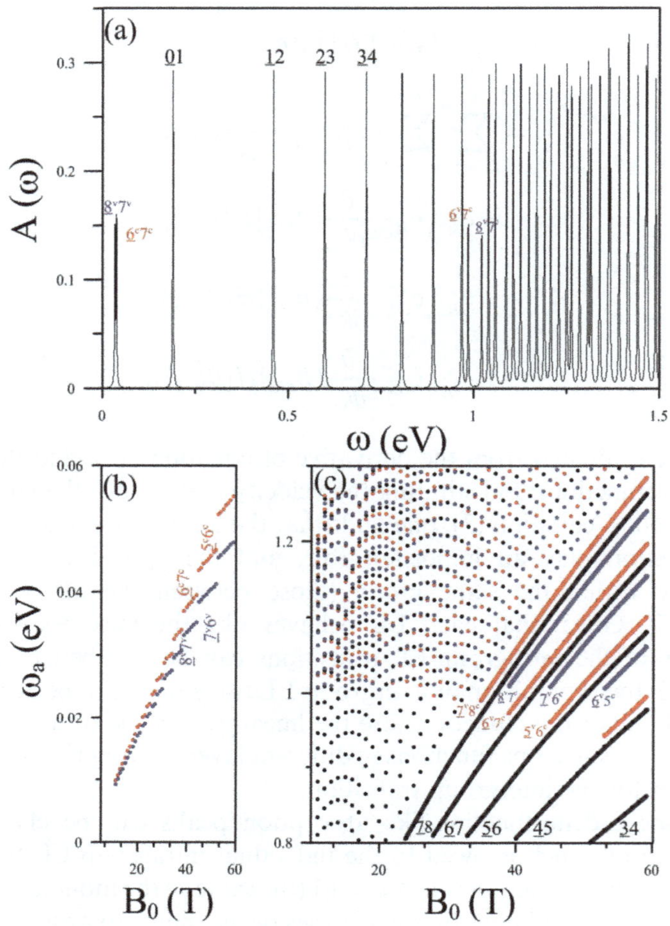

Figure 3.6. (a) The magneto-absorption spectrum of AA-stacked trilayer graphene under B_0 = 40 T. The three kinds of intragroup excitation channels are indicated by different colors. The B_0-dependent frequencies of the threshold and full channels are shown in (b) and (c), respectively.

Dirac cones. However, it should be noted that for the four peaks neighboring the forbidden transition region, $\omega_1^{8^v7^v}$, $\omega_3^{6^c7^c}$, $\omega_1^{8^v7^c}$ and $\omega_3^{6^v7^c}$, their intensities are about half of the others due to only one kind of channel being available for the excitations from the Fermi level.

It is useful to elaborate the magnetic quantization of AA-stacked graphene in terms of the numbers and frequencies of the field-dependent absorption peaks in the three categories, as in figure 3.6(b). During the variation of the magnetic field, the Dirac fermion excitations behave as a direct consequence of the $\sqrt{B_0}$ proportionality of the absorption frequency. When the Fermi level crosses two LLs in the bilayer-like groups, a discontinuity of the threshold frequency appears accompanied by the change of the threshold channel in the $\omega - B_0$ spectrum. As B_0 is reduced, the threshold frequencies decrease in the first and third categories and they are dominated by larger quantum numbers, such as $\omega_1^{8^v7^v}$ and $\omega_3^{6^c7^c}$ at $B_0 = 40$ T and

$\omega_1^{7^v6^v}$ and $\omega_3^{5^c6^c}$ at $B_0 = 50$ T. The main reason for this is that the threshold frequency contributed by the excitations from the highest occupied LL is strongly affected by the LL degeneracy, which is proportional to the magnetic field strength. That is to say, at a constant carrier density, the weaker the magnetic field is, the more the valence LLs are occupied. In addition, the forbidden transition region also depends on the magnetic field strength. It converges on the frequency at approximately $2\alpha_1 \simeq 0.73$ eV, corresponding to the difference between the two outer Dirac points (figure 3.1). With an increase of the graphene sheets, we can classify the spectra depending on whether the numbers are even or odd. All the categories of absorption peaks in the even-layer scenario can be regarded as bilayer-like spectra, each with a discontinuous threshold frequency and a forbidden transition region. Except for a monolayer-like spectrum exclusively shown for the odd-layer scenario, all the others are bilayer-like.

3.3 Electric-field-enriched optical properties

By means of applying a perpendicular electric field $\mathbf{E} = E_0\hat{z}$, one can modify the spectral features, such as those from Dirac fermions, saddle-point states and Landau states. A uniform perpendicular electric field does not break the linear dispersions but only causes a rigid shift of the Dirac cone structures with a slight decrease of the linear slope. The upper and lower Dirac cones shift in opposite directions corresponding to the middle one that is placed at the Fermi level. The energy shift is increased with an increment of the electric field strength [13]. The spectral intensity is also linearly dependent on the frequency, while it slightly increases due to the slope variation for the Dirac-type linear dispersion, as shown in figure 3.7. However, the shoulder structure, at the doubled frequency between two Dirac points, exhibits a blue shift in the same magnitude with respect to the enlargement of the forbidden transition zone across the Fermi level.

Responsible for the Dirac cone shift, the first and third groups of LLs are also rigidly shifted to opposite directions with regard to the second group. The selection rule is the same as the case without an electric field, because the quantum modes of the Landau states are unchanged and still described by well-behaved Landau wavefunctions. Consequently, the second category of absorption peaks is almost unchanged under the influence of the electric field, while for the other two categories, the forbidden transition region is enlarged, with the threshold frequencies and channels dominated by larger quantum numbers. In short, the applied electric field rarely changes the main characteristics of the absorption spectrum because it does not break the spatial symmetry in the AA stacking configuration.

The aforementioned absorption spectral features can be verified by optical spectroscopies, such as absorption, transmission, reflection, Raman scattering and Rayleigh scattering spectroscopies. The multiple Dirac cone dispersion is verified by ARPES in the bilayer case [21], while the optical excitation spectrum for multilayer cases is still absent. On the other hand, the previous experiments have identified the $\sqrt{B_0}$- dependent frequencies in the infrared transmission spectra for monolayer and AB-stacked trilayer graphene (chapter 4). Recently, this has also been observed by

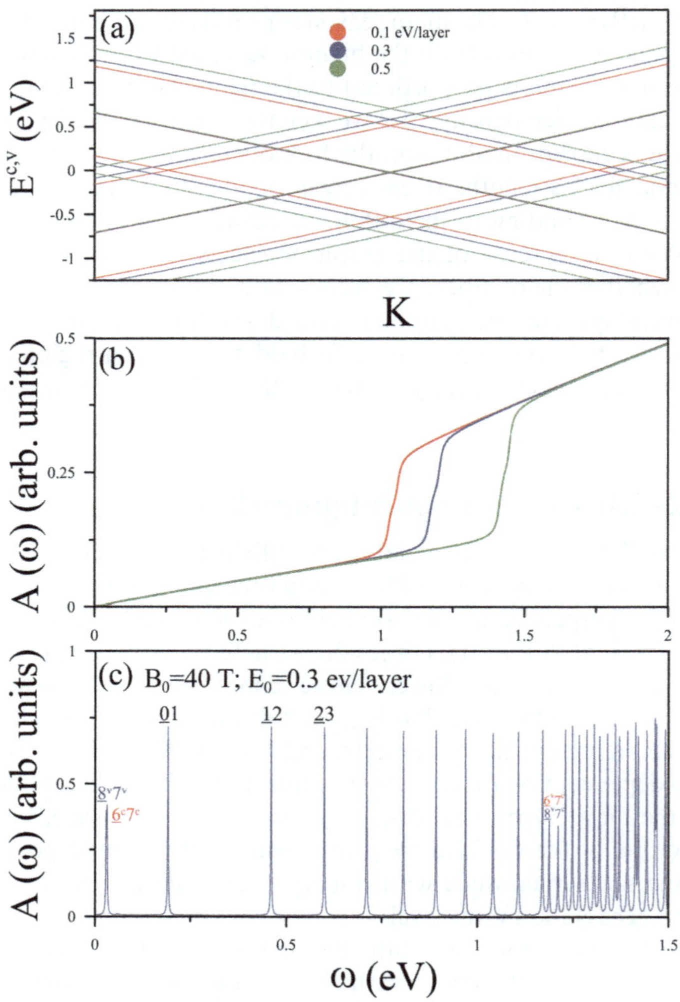

Figure 3.7. The E_0-enriched (a) band structure, (b) absorption spectrum and (c) magneto-absorption spectrum of the AA-stacked trilayer graphene.

magneto-Raman spectroscopy for AB-stacked graphene up to five layers [22]. While the magneto-optical properties of massless Dirac fermions have been observed, verifying the multiple Dirac points in the AA stacking configuration would be worthwhile in further research to identify interlayer atomic interactions and geometric structures.

References

[1] Lee J K, Lee S C, Ahn J P, Kim S C, Wilson J I B and John P 2008 The growth of AA graphite on (1 1 1) diamond *J. Chem. Phys.* **129** 234709

[2] Lobato I and Partoens B 2011 Multiple Dirac particles in AA-stacked graphite and multilayers of graphene *Phys. Rev.* B **83** 165429

[3] Shyu F L and Lin M F 2001 Low-frequency π-electronic excitations of simple hexagonal graphite *J. Phys. Soc. Jpn* **70** 897–901

[4] Lin M F, Huang C S and Chuu D S 1997 Plasmons in graphite and stage-1 graphite intercalation compounds *Phys. Rev.* B **55** 13961

[5] Lin M F, Chuu D S, Huang C S, Lin Y K and Shung K W K 1996 Collective excitations in a single-layer carbon nanotube *Phys. Rev.* B **53** 15493

[6] Lin M F, Shyu F L and Chen R B 2000 Optical properties of well-aligned multiwalled carbon nanotube bundles *Phys. Rev* B **61** 14114

[7] Watzel J, Pavlyukh Y, Schaffer A and Berakdar J 2016 Optical vortex driven charge current loop and optomagnetism in fullerenes *Carbon* **99** 439–43

[8] Charlier J C, Michenaud J P and Gonze X 1992 First-principles study of the electronic properties of simple hexagonal graphite *Phys. Rev.* B **46** 4531–9

[9] Chang C P 2011 Exact solution of the spectrum and magneto-optics of multilayer hexagonal graphene *J. Appl. Phys.* **110** 013725

[10] Kim K S, Walter A L, Moreschini L, Seyller T, Horn K and Rotenberg E *et al* 2013 Coexisting massive and massless Dirac fermions in symmetry-broken bilayer graphene *Nat. Mater.* **12** 887–92

[11] Ho J H, Lu C L, Hwang C C, Chang C P and Lin M F 2006 Coulomb excitations in AA- and AB-stacked bilayer graphites *Phys. Rev.* B **74** 085406

[12] Chiu C W, Chen S C, Huang Y C, Shyu F L and Lin M F 2013 Critical optical properties of AA-stacked multilayer graphenes *Appl. Phys. Lett.* **103** 041907

[13] Tsai S J, Chiu Y H, Ho Y H and Lin M F 2012 Gate-voltage-dependent Landau levels in AA-stacked bilayer graphene *Chem. Phys. Lett.* **550** 104–10

[14] Gerhardts R R, Weiss D and Vonklitzing K 1989 Novel magnetoresistance oscillations in a periodically modulated two-dimensional electron gas *Phys. Rev. Lett.* **62** 1173

[15] Terashima T, Kikugawa N, Kiswandhi A, Choi E S, Brooks J S and Kasahara S *et al* 2014 Anomalous Fermi surface in FeSe seen by Shubnikov-de Haas oscillation measurements *Phys. Rev.* B **90** 144517

[16] Zudov M A, Du R R, Simmons J A and Reno J L 2013 Shubnikov-de Haas-like oscillations in millimeterwave photoconductivity in a high-mobility two-dimensional electron gas *Phys. Rev.* B **64** 201311

[17] Ho Y H, Wu J Y, Chen R B and Lin M F 2010 Optical transitions between Landau levels: AA-stacked bilayer graphene *App. Phys. Lett.* **97** 101905

[18] Koshino M and Ando T 2008 Magneto-optical properties of multilayer graphene *Phys. Rev.* B **77** 115313

[19] Abergel D S L and Fal'ko V I 2007 Optical and magneto-optical far-infrared properties of bilayer graphene *Phys. Rev.* B **75** 155430

[20] Ho Y H, Chiu Y H, Lin D H, Chang C P and Lin M F 2010 Magneto-optical selection rules in bilayer Bernal graphene *ACS Nano* **4** 1465–72

[21] Kim K S, Walter A L, Moreschini L, Seyller T, Horn K and Rotenberg E *et al* 2013 Coexisting massive and massless Dirac fermions in symmetry-broken bilayer graphene *Nat. Mat.* **12** 887–92

[22] Berciaud S, Potemski M and Faugeras C 2014 Probing electronic excitations in mono- to pentalayer graphene by micro magneto-Raman spectroscopy *Nano Lett.* **14** 4548–53

Chapter 4

AB-stacked graphene

AB stacking is known as the most stable and common stacking configuration among trilayer graphene systems, as predicted by the theoretical studies [1–4] and confirmed by the experimental measurements [5–9]. It is analogous to the composite of monolayer and AB-stacked bilayer graphene. The energy band structure is composed of a pair of linear subbands and two pairs of parabolic subbands near the K point. There are three groups of valence and conduction LLs, in which the monolayer- and bilayer-like varieties, respectively, have an almost symmetric and asymmetric energy spectrum about the Fermi level. Each one is four-fold degenerate because of the extremely small splitting energy, except for the three special double-degenerate LLs, $n_1^{c,v} = 0$ and $n_2^{c,v} = 0$ and 1.

Most LL wavefunctions belong to the well-behaved modes, and the perturbed ones only appear in a few anticrossings. The zero-field absorption spectrum mainly exhibits two kinds of special structure, including discontinuities (shoulders after broadening) at low frequencies and logarithmically divergent peaks at middle frequencies. There are only five kinds of excitation channel due to the forbidden transitions between the linear and parabolic energy bands. Up to now, infrared reflection [10–13], absorption spectroscopies [14–16] and infrared scattering scanning near-field optical microscopy [17] have successfully verified the characteristic structure and the AB stacking domains.

A perpendicular electric field can dramatically change electronic structures and thus diversify optical properties [7, 18–24]. This significantly modulates the subband spacings, causes subband hybridizations and induces extra band-edge states [18]. One pair of parabolic subbands near E_F becomes oscillatory with three circular constant energy contours in the energy wave vector space. The second pair might have sombrero-shaped dispersions with two constant energy loops. These will lead to asymmetric absorption peaks in the square root divergent form. Moreover, it is expected that the increase in the band-edge states enhances the density of states, creating extra excitation channels and enriching the spectral structures.

doi:10.1088/978-0-7503-1566-1ch4 © IOP Publishing Ltd 2017

Furthermore, a perpendicular magnetic field can generate plentiful absorption spectra [25, 26]. The five categories of inter-LL excitations differ quite significantly from one another in intensity, frequency and structure of absorption peaks. Moreover, the magneto-absorption spectrum can be diversified by the difference in Coulomb potential energy between graphene layers. The four-fold degenerate LLs become double-degenerate, so that more inter-LL channels are excited. The spectral intensity is lower with the increase of absorption peaks and the appearance of double-peak structures.

4.1 Electronic and optical properties without external fields

The geometric structure of the AB-stacked configuration is shown in Figure 2.1(c), where two neighboring layers shift relative to each other by one bond length b along the armchair direction. The A atoms on the odd-number layers and the B atoms on the even-number layers have the same xy projections, while the B (A) atoms on the odd- (even-) number layers are located at the hexagon centers of the adjacent layer. The primitive unit cell consists of $2 \times N$ atoms, where N is the number of layers. The atomic interactions are derived from the SWMcC model [27, 28], including one nearest-neighbor intralayer γ_0, three nearest-neighbor interlayers (γ_1, γ_3, γ_4), two next-nearest-neighbor interlayers (γ_2, γ_5), and the chemical environment difference γ_6 between A and B atoms. Their values are listed as follows: $\gamma_0 = -3.12$ eV, $\gamma_1 = 0.38$ eV, $\gamma_2 = -0.021$ eV, $\gamma_3 = 0.28$ eV, $\gamma_4 = 0.12$ eV, $\gamma_5 = -0.003$ eV, and $\gamma_6 = -0.0366$ eV [27, 28].

The electronic properties in AB-stacked trilayer graphene can be basically regarded as a hybridization of monolayer and bilayer properties, while the interlayer atomic interactions play an important role on the energy dispersions in the vicinity of highly symmetric points, as indicated in figure 4.1(a). The low-energy band structure is characterized by massless and massive Dirac fermions; it consists of a pair of linear subbands, labeled as $S_1^{c,v}$, and two pairs of parabolic subbands, $S_2^{c,v}$ and $S_3^{c,v}$, in the vicinity of the K point. Analytic solutions show that γ_3 and γ_4 modify the low-lying energy dispersions, and γ_2 and γ_5 separate the intersections of the nearby subbands and influence their spacings as well [18, 29]. Nevertheless, a slight overlap between the linear valence band and that of the parabolic conduction indicates a gapless 2D semimetal.

Away from $E_F = 0$, the two subbands $S_3^{c,v}$ exhibit parabolic dispersions and they are separated mainly by γ_1, with a fluctuation of γ_2, γ_5 and γ_6. Furthermore, these parabolic dispersions extend along the KM direction, with three saddle-points near the M point. It is remarkable that the energy dispersions near the K and M points give rise to special structures in the DOS [18, 29] and dominate the main characteristics of the absorption spectra. Recently, the theoretical band structure calculations were confirmed by STS and ARPES measurements [5, 23, 30].

In the absence of an electric field, the absorption spectrum of trilayer graphene mainly exhibits two kinds of special structure: discontinuities at low frequencies and logarithmic divergences at middle frequencies, which, respectively, are owed to the vertical transitions around the K point and those around the M point (Figure 4.1(b)). The 3×3 excitation channels are induced between three pairs of conduction

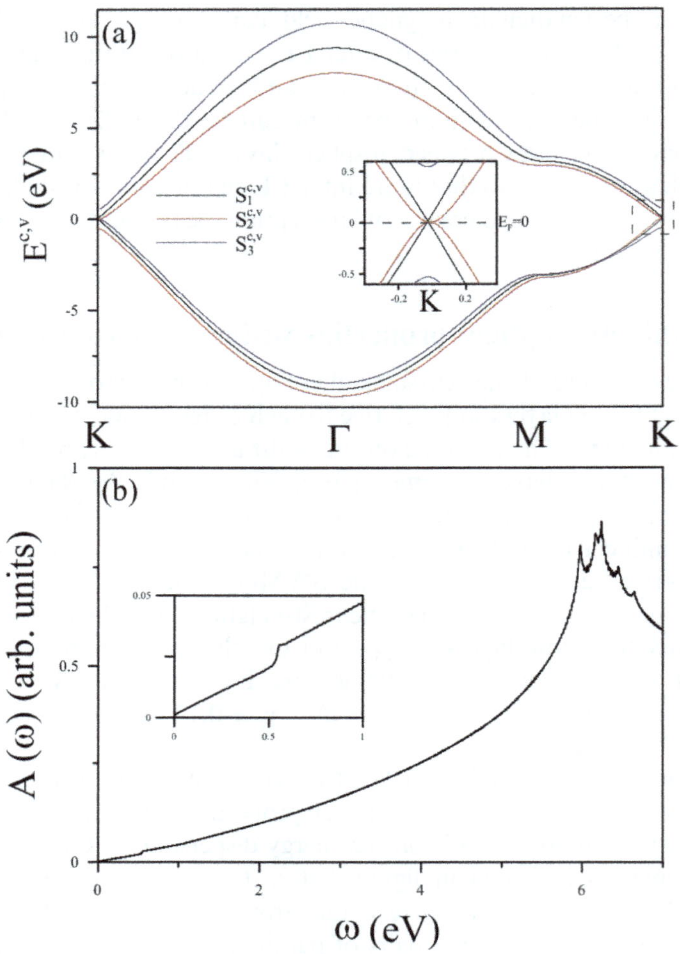

Figure 4.1. (a) The zero-field band structure and (b) absorption spectrum of AB-stacked trilayer graphene. In (a), the three subbands are labeled as $S_1^{c,v}$ (black), $S_2^{c,v}$ (red) and $S_3^{c,v}$ (blue). The insets of (a) and (b) show the zoomed-in view at low energies.

and valence subbands; however, on the basis of the symmetry characteristics of the subenvelope functions, the permitted channels are only those completely within the monolayer-like subbands $S_1^{c,v}$ or the bilayer-like subbands $S_2^{c,v}$ and $S_3^{c,v}$. That is to say, there are only five kinds of channel that are accessible in the absorption spectra, while noticeably, the optical excitations are forbidden from $S_1^{c,v}$ to $S_2^{c,v}$ or $S_3^{c,v}$.

At low frequencies, the spectral structure is determined by the latter excitation channels, since those between the linear bands only contribute to the linearly increasing intensity. The special shoulder structure at $\omega \sim 0.5$ eV is due to the edge-state transitions from $S_2^v \rightarrow S_3^c$ and $S_3^v \rightarrow S_2^c$, but otherwise the spectral intensity grows with the higher frequency. In the middle-frequency spectrum, there are five logarithmic peaks around $\omega = 2\gamma_0$, coming from the saddle-points of the $S_1^{c,v}$, $S_2^{c,v}$ and $S_3^{c,v}$ subbands. Recently, infrared reflection spectroscopy and absorption spectroscopy showed clear evidence of the characteristic structure at $\omega \sim 0.5$ eV [10, 12]. Furthermore, infrared scattering scanning near-field optical microscopy can be used to verify the AB stacking domains based on specific conductivities [17]. The whole frequency spectrum deserves closer experimental examination to verify the whole band structure and to identify the stacking configuration [9].

4.2 Magneto-electronic and magneto-optical properties

The magnetic Hamiltonian matrix elements of the ABA stacking are expressed as

$$\begin{aligned}
\left\langle A_m^L|H|A_{m'}^{L'}\right\rangle &= \gamma_1\delta_{m,m'}\delta_{L,L'\pm1} + (\gamma_5+\gamma_6)\delta_{m,m'}\delta_{L,L'} \\
&\quad + \frac{\gamma_5}{2}\delta_{m,m'}\delta_{L,L'\pm2}, \\
\left\langle B_m^L|H|B_{m'}^{L'}\right\rangle &= \gamma_3\left(t_{2,k}\delta_{m,m'} + t_{4,k}^*\delta_{m,m'-1}\right)\delta_{L,L'\pm1} + \gamma_2\delta_{m,m'}\delta_{L,L'} \\
&\quad + \frac{\gamma_2}{2}\delta_{m,m'}\delta_{L,L'\pm2}, \\
\left\langle A_m^L|H|B_{m'}^{L'}\right\rangle &= \gamma_0\left(t_{4,k}\delta_{m,m'} + t_{3,k}^*\delta_{m,m'+1}\right)\delta_{L,L'} \\
&\quad + \gamma_4\left(t_{1,k}\delta_{m,m'} + t_{4,k}^*\delta_{m,m'+1}\right)\delta_{L,L'\pm1} \text{ for odd L}, \\
&= \gamma_0\left(t_{1,k}\delta_{m,m'} + t_{4,k}^*\delta_{m,m'+1}\right)\delta_{L,L'} \\
&\quad + \gamma_4\left(t_{4,k}\delta_{m,m'} + t_{3,k}^*\delta_{m,m'+1}\right)\delta_{L,L'\pm1} \text{ for even L}.
\end{aligned} \tag{4.1}$$

ABA stacking exhibits three unique groups of LLs, which can be regarded as the superposition of those of monolayer and AB-stacked bilayer graphene, as indicated in figure 4.2(a)–(g). The first group of LLs, being quantized from the linear subbands, is initially formed near E_F. In the vicinity of E_F, the square root dependence of LL energies on $n_1^{c,v}$ is absent because of the slightly distorted Dirac cone structure with an energy spacing of ~20 meV. Furthermore, the $n_1^{c,v} = 0$ LLs are doubly degenerate and their energies are different at the 2/6 and 1/6 (5/6 and 4/6) centers. The $n_1^{c,v} \geq 1$ LLs, whose conduction and valence bands are almost symmetric to each other, are four-fold degenerate, similar to those of monolayer graphene. The second and third groups come from the two pairs of parabolic subbands. The $n_2^{c,v}$ LLs are initiated at $E^{c,v} \approx 0$, while the $n_3^{c,v}$ LLs are located at (0.57 eV, −0.6 eV). Each LL is double-degenerate, mainly owing to the destruction of the mirror symmetry about the z-plane. The energy spacing between the splitting LLs is extremely small (≈0.1 meV), so that their absorption peaks will merge together. Exceptionally, the $n_2^{c,v} = 0$ and 1 LLs exhibit observable splitting energies of about 10 meV, accompanied by certain important differences in the LL wave-functions and magneto-absorption spectra at the 2/6 and 1/6 centers (5/6 and 4/6 centers). Moreover, all the LLs possess well-behaved spatial distributions except for a few anticrossings in the B_0-dependent energy spectrum. It is also noted that the low-lying LLs of the first and second groups have been identified from the STS [31] and the quantum Hall transport measurements [32, 33].

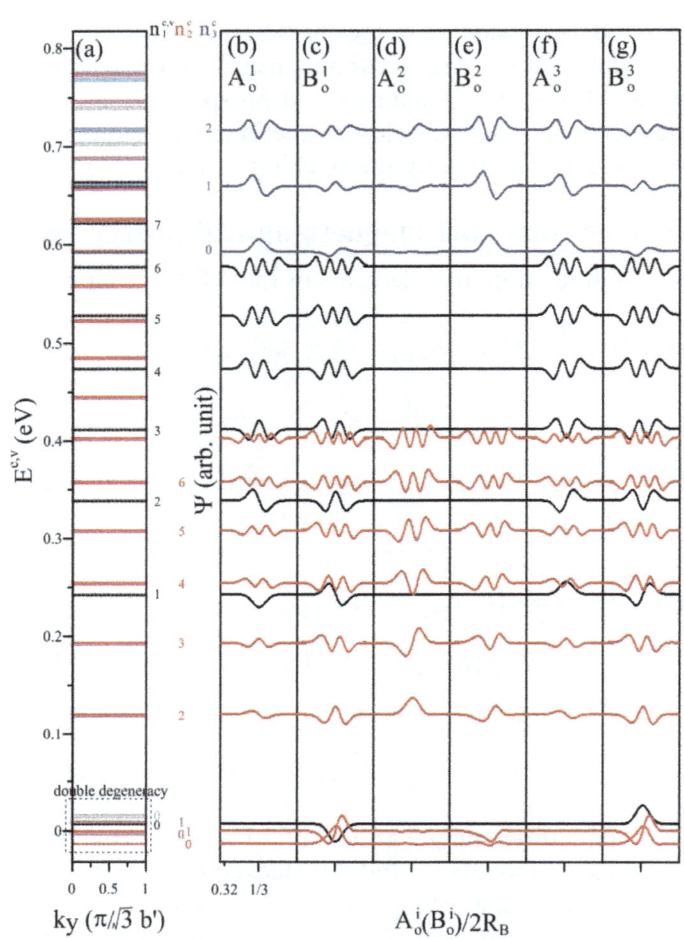

Figure 4.2. The Landau levels and subenvelope functions of 2/6 localized Landau states for AB-stacked trilayer graphene at $B_0 = 40$ T. The energies of 1/6 localized Landau states are plotted in (a) for comparison (light colors).

There are five categories of absorption peaks in the optical spectrum, as observed in the zero-field available excitation channels (figure 4.3). The intergroup optical excitations associated with the first LL group are forbidden due to the simple phase relations among six different sublattices. The first group only survives on (A^1, B^1, A^3, B^3) sublattices, since the neighboring-layer atomic interaction does not make a contribution to its electronic states. Furthermore, the A^1 and A^3 (B^1 and B^3) subenvelope functions are identical for the second and third groups, whereas those of the first group possess the opposite phases. As to the intragroup excitations $n_1^v \rightarrow n_1^c$, they exhibit only single peaks with a predominantly uniform intensity because the velocity matrix elements are nearly constant for the almost symmetric Dirac cone structure.

The second and third LL groups sharply contrast with the first one in terms of intensity, structure and frequency of intragroup absorption peaks. The peak

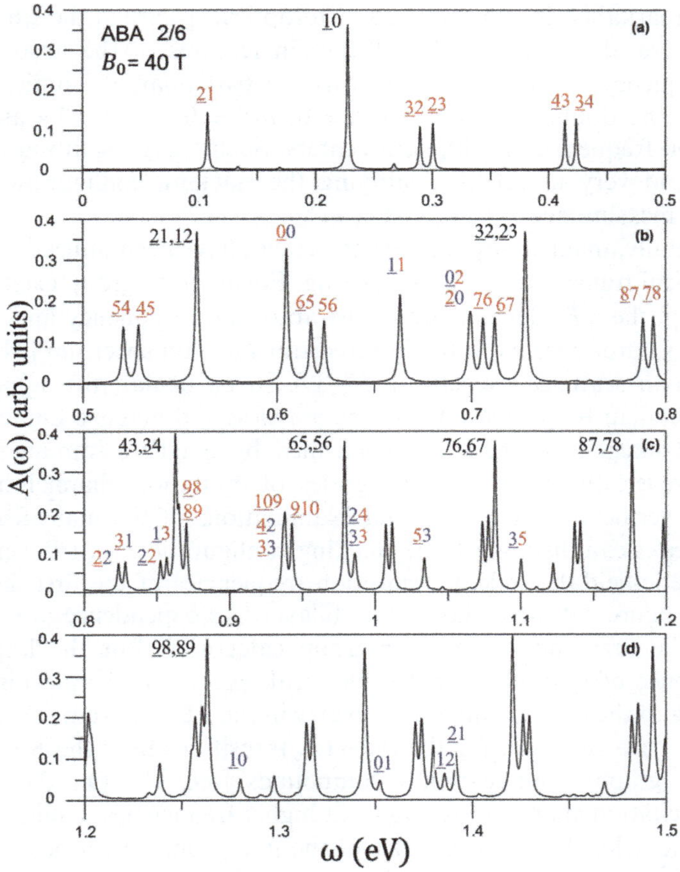

Figure 4.3. The magneto-absorption spectrum of AB-stacked trilayer graphene with three intragroup and two intergroup categories of LL excitations under B_0 = 40 T. The black, red and blue numbers, respectively, indicate the excitations from the first, second and third LL groups.

intensities are non-uniform and only about half of that of the first group. The magneto-absorption spectrum presents twin-peak structures with an observable spacing of ~10 meV. The differences in peak structures can account for the lower intensity. Exceptionally, the $\omega_2^{2^v1^c}$ peak is a single structure because of the requirement of the Fermi–Dirac distribution. Concerning the two categories of intergroup excitations, $n_2^v \to n_3^c$ and $n_3^v \to n_2^c$ display the single peaks. Such peaks obey the selection rules of $|n_2^{c,v} - n_3^{v,c}| = 2$ and 0, and their intensities are comparable. In general, the intensity of the intergroup inter-LL excitations is weaker than half of that of the intragroup ones. This is mainly due to the phase relations between the A^l (B^l) subenvelope functions of the initial states and B^l (A^l) of the final ones. For the intragroup excitations, all the subenvelope functions of the n^v initial and n^c final states on the A^l and B^l sublattices have the same phases (for the second group) or the opposite (for the third one), resulting in high peak intensities. On the other hand, the complex phase relations in the wavefunctions weaken the intergroup inter-LL transitions remarkably. In short, the magneto-optical spectra at the 2/6 (5/6) and 1/6 (4/6) centers are almost identical in the main features of the absorption peaks, including frequency, structure and intensity. In particular, the optical excitations coming from the double-degenerate $n_1^{c,v} = 0$, $n_2^{c,v} = 0$ and 1 LLs are distinct in the absorption frequencies at different centers. Such energy splitting is observable (~10 meV) and very useful in identifying the stacking configurations from the experimental measurements.

Experimentally, magneto-optical spectroscopies have been utilized to observe the inter-LL vertical transitions in ABA stacking. For the intragroup excitations of the first LL group, the $\sqrt{B_0}$-dependence of the absorption frequency and the selection rule of $\Delta n = \pm 1$ are revealed in the infrared transmission spectrum [34–36] and the magneto-Raman scattering spectrum [37–39]. In particular, the square root and linear B_0-dependent frequencies due to the massless and massive Dirac fermions of few-layer AB-stacked graphene are confirmed by magneto-Raman spectroscopy [37–39]. However, the other four categories of excitation channels need further experimental verification. Experimental examinations of the main features of the absorption peaks can distinguish the stacking configurations of trilayer graphene.

The magnetic-field-dependent absorption frequencies of the first few categories are shown in figure 4.4. Monolayer- and bilayer-like dependence are, respectively, reflected in the first and second intragroup categories. For the latter, the tiny energy difference of ~10 meV for the twin-peak excitation channels is revealed as corresponding to the electron–hole asymmetry in the LL spectrum [18–29]. Splitting energies have been observed for the inter-LL transitions near the K point in bulk graphite and graphite intercalation compounds [40, 41]. On the other hand, intergroup excitation channels are shown at higher frequencies. Under a sufficiently small magnetic field, the onset energies of the intergroup categories correspond to the band-edge states in figure 4.1, which are mainly affected by the interaction, γ_1. The aforementioned results imply that optical experiments can be used to determine the interlayer atomic interactions that dominate the electron–hole asymmetry and band dispersions through the observation of the splitting energies in the twin-peak structure and the convergent threshold frequencies of the intergroup categories.

4.3 Electric-field-enriched optical properties

A perpendicular electric field significantly modulates the low-lying energy dispersions and optical excitation channels. It changes subband spacings, causes subband hybridizations and induces extra band-edge states as well (Figure 4.5(a)). Unlike the induction of a significant band gap in bilayer graphene, trilayer graphene has been shown both theoretically [42] and experimentally [6] to behave as a semimetal with weakly oscillating dispersions near the Fermi level. Attributed to the hybridization of the linear and parabolic subbands, the oscillation width is greatly enhanced as the applied electric field increases. In particular, the pair of parabolic subbands near E_F becomes oscillatory, similar to the $S_1^{c,v}$ subbands in AAB stacking (figure 6.1 in section 6.1). In the energy wave vector space, each subband has three circular constant energy contours associated with the concave and convex dispersions.

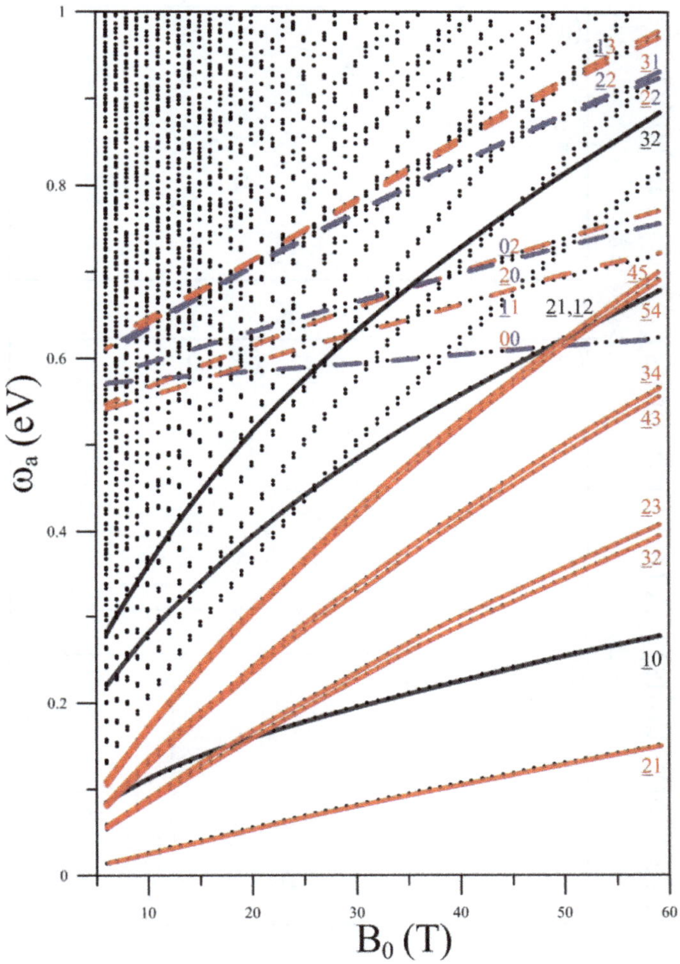

Figure 4.4. The B_0-dependent magneto-absorption frequencies of AB-stacked trilayer graphene. The solid and dashed curves indicate the excitation channels of the intragroup and intergroup categories, respectively.

This will lead to asymmetric absorption peaks in the square root divergence form, since such subbands can be regarded as one-dimensional. Similar behavior can be observed in sombrero-shaped bands with two constant energy loops. Moreover, it is expected that the increase in the band-edge states of the oscillating subbands enhances the density of states, induces optical excitation channels and diversifies the low-energy optical spectra.

The characteristics of the absorption spectra, such as the prominent structures, number, intensity and frequency of absorption peaks, change significantly under the influence of the applied electric field (Figure 4.5(b)). Owing to the subband hybridizations, not only do the original transition channels survive, but extra channels are also created between the first pair and the other two pairs of energy bands. That is to say, five kinds of effective excitation channels are changed into nine under the application of **E**. The low-energy optical excitation channels from the

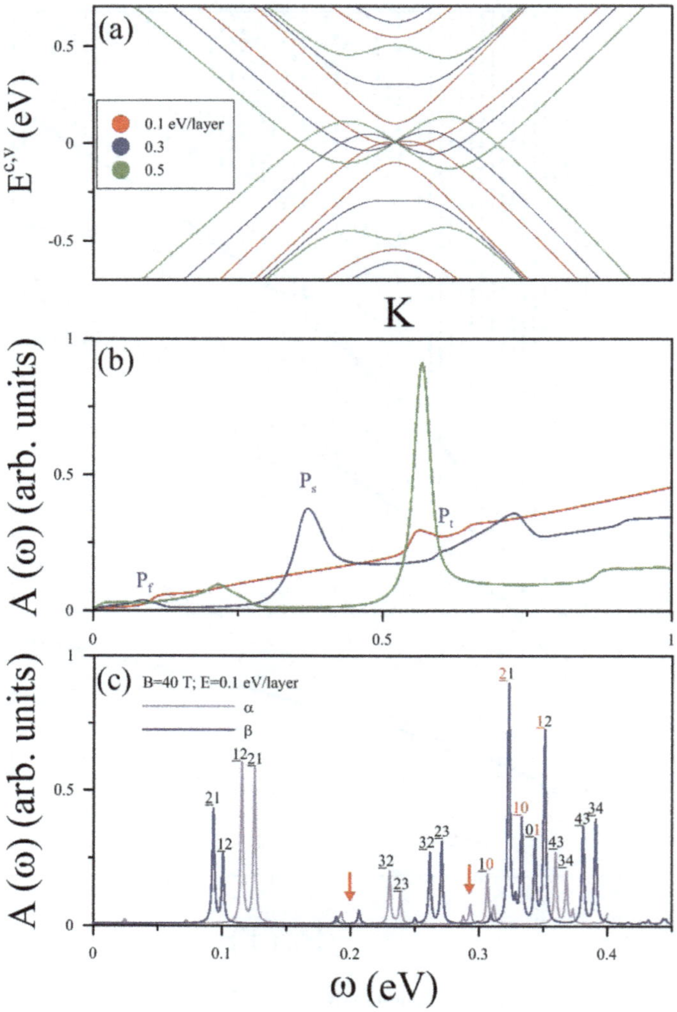

Figure 4.5. The E_0-enriched (a) band structure, (b) absorption spectrum and (c) magneto-absorption spectrum of AB-stacked trilayer graphene. The α- and β-type spectra, respectively, stand for the excitations from the two valley-splitting LLs, 1/6 and 2/6 localized states in figure 4.2.

field-induced oscillatory subbands give rise to multipeak structures in the absorption spectra due to the multiple channels, instead of the featureless spectra without external fields.

Accordingly, as the field strength is larger than a moderate value ~0.3 eV/layer, three multipeak structures appear in the low-frequency region, as shown in figure 4.5(b). The first, second and third peaks, as denoted by P_f, P_s and P_t, correspond to the transitions from $S_1^v \rightarrow S_1^c$, ($S_1^v \rightarrow S_2^c$, $S_2^v \rightarrow S_1^c$), $S_2^v \rightarrow S_2^c$, respectively. The number of excitation channels for each multipeak depends on the number of band-edge states of the oscillatory subbands; that is, it depends on the electric field strength. At E_0 = 0.3 eV/layer (red curve), each multipeak structure consists of one main peak accompanied by one or two weak peaks (like shoulders after broadening). However, the absorption spectra exhibit more and sharper symmetric peaks with an increment of the field strength, especially for P_s, which displays a rather strong symmetric peak as E_0 goes up to 0.5 eV/layer (green curve). The peak intensity is enhanced by the larger radius of the constant energy loop (the higher density of states). Specifically, the symmetric peak structure of P_s is a combination of two asymmetric peaks in opposite directions due to the concave- and convex-downward subbands ($S_1^v \rightarrow S_2^c$, $S_2^v \rightarrow S_1^c$) in figure 4.5(a). These results are in qualitative agreement with recent transport [42] and optical experiments [7]. Furthermore, the excitation energy of the first peak P_f is nearly independent of the electric field strength, which is responsible for the semimetal properties in the trilayer case of AB stacking. The other peaks show blue shifts with an increasing field strength. It should be noted that the **E**-induced low-frequency absorption peaks rely on the geometric structure. By contrast, the characteristics of the middle-frequency spectra due to the π-electronic states around the M point are preserved in the presence of an electric field.

The applied perpendicular electric field can dramatically alter the LL characteristics and thus diversify the magneto-absorption spectrum (Figure 4.5(c)). The fourfold degenerate LLs are replaced by double-degenerate LLs under broken mirror symmetry about the z-plane. Each group of LLs is split into two subgroups, in which the splitting LLs are characterized by two different localization centers of the 1/6 and 2/6 unit cell (4/6 and 5/6 unit cell). The dominating sublattice, in determining the quantum numbers, remains unchanged for each subgroup, i.e. there exist two inversely symmetric sublattices for two splitting subgroups. As more inter-LL channels are excited, the absorption spectrum shows double peaks, identified as α and β types for the inter-LL transitions from the former and the latter subgroups, respectively. Furthermore, anticrossings of the perturbed LLs are presented in the E_0-dependent energy spectrum (red arrow), since the cooperation of the electric field and certain interlayer atomic interactions can drastically change the charge distributions in the main and side modes (details discussed in section 5.2). The total number of absorption peaks is greatly increased, but their intensities are significantly reduced. The double and the twin peaks, which arise from the splitting and asymmetric LL energy spectra respectively, are simultaneously revealed in the magneto-optical excitations. The two components in each double peak are associated with the 2/6 and 1/6 centers, while those in the twin-peak structure only rely on a specific center. It should be noted that the inter-LL excitations between the first group and the others are recovered in the presence of **E**. The potential energy differences between the graphene layers result in drastic changes of the spatial charge distributions; that is, the first group of LLs is contributed by the electronic states in each layer. The existence of second-layer charge distribution is responsible for the enriched absorption peaks. The electric field can generate nine categories of magneto-absorption peaks, which are absent in the AAA, ABC and AAB stacking systems. Also, there exist extra absorption peaks arising from the frequent intra-subgroup LL anticrossings. The **E**-induced intergroup excitation channels and extra peaks are expected to provide more information in identifying the stacking domains from the magneto-optical measurements [25, 25, 43, 44].

The double-degenerate splitting LL energy spectrum in the ABA stacking was identified from the experimental measurements on the Hall conductances [44] under a composite field. The **E**-created diverse magneto-optical properties can be further verified by infrared transmission spectroscopy [36] and Raman scattering spectroscopy [45], in which they include nine categories of inter-LL excitation channels, the anticrossing-dependent extra peaks, the simultaneous appearance of the single, twin and double peaks and the weakened peak intensities.

References

[1] Min H and MacDonald A H 2008 Electronic structure of multilayer graphene *Prog. Theor. Phys. Supplement* **176** 227–52

[2] Gruneis A, Attaccalite C, Wirtz L, Shiozawa H, Saito R and Pichler T *et al* 2008 Tight-binding description of the quasiparticle dispersion of graphite and few-layer graphene *Phys. Rev.* B **78** 205425

[3] Aoki M and Amawash H 2007 Dependence of band structures on stacking and field in layered graphene *Solid State Commun.* **142** 123–7

[4] Avetisyan A A, Partoens B and Peeters F M 2010 Stacking order dependent electric field tuning of the band gap in graphene multilayers *Phys. Rev.* B **81** 115432

[5] Bao W, Jing L, Velasco J, Lee Y, Liu G and Tran D *et al* 2011 Stacking-dependent band gap and quantum transport in trilayer graphene *Nat. Phys.* **7** 948–52

[6] Coletti C, Forti S, Principi A, Emtsev K V, Zakharov A A and Daniels K M *et al* 2013 Revealing the electronic band structure of trilayer graphene on SiC: an angle-resolved photoemission study *Phys. Rev.* B **88** 155439

[7] Lui C H, Li Z, Mak K F, Cappelluti E and Heinz T F 2102 Observation of an electrically tunable band gap in trilayer graphene *Nat. Phys.* **7** 944–7

[8] Ohta T, Bostwick A and McChesney J L 2007 Interlayer interaction and electronic screening in multilayer graphene investigated with angle-resolved photoemission spectroscopy *Phys. Rev. Lett.* **98** 206802

[9] Mak K F, Shan J and Heinz T F 2010 Electronic structure of few-layer graphene: experimental demonstration of strong dependence on stacking sequence *Phys. Rev. Lett.* **104** 176404

[10] Zhang L M, Li Z Q, Basov D N, Fogler M M, Hao Z and Martin M C 2008 Determination of the electronic structure of bilayer graphene from infrared spectroscopy *Phys. Rev.* B **78** 235408

[11] Li Z Q, Henriksen E A, Jiang Z, Hao Z, Martin M C and Kim P *et al* 2009 Band structure asymmetry of bilayer graphene revealed by infrared spectroscopy *Phys. Rev. Lett.* **102** 037403

[12] Kuzmenko A B, van Heumen E, van der Marel D, Lerch P, Blake P and Novoselov K S *et al* 2009 Infrared spectroscopy of electronic bands in bilayer graphene *Phys. Rev.* B **79** 115441

[13] Berciaud S, Potemski M and Faugeras C 2014 Probing electronic excitations in mono- to pentalayer graphene by micro magneto-raman spectroscopy *Nano Lett.* **14** 4548–53

[14] Zhang Y, Tang T T, Girit C, Hao Z, Martin M C and Zettl A *et al* 2009 Direct observation of a widely tunable bandgap in bilayer graphene *Nature* **459** 820–3

[15] Li Z Q, Henriksen E A, Jiang Z, Hao Z, Martin M C and Kim P *et al* 2008 Dirac charge dynamics in graphene by infrared spectroscopy *Nat. Phys.* **4** 532–5

[16] Wang F, Zhang Y, Tian C, Girit C, Zettl A and Crommie M *et al* 2008 Gate-variable optical transitions in graphene *Science* **320** 206–9

[17] Kim D S, Kwon H, Nikitin A Y, Ahn S, Martin-Moreno L and Garcia-Vidal F J *et al* 2015 Stacking structures of few-layer graphene revealed by phase-sensitive infrared nanoscopy *ACS Nano* **9** 6765–73

[18] Lu C L, Chang C P, Huang Y C, Chen R B and Lin M L 2006 Influence of an electric field on the optical properties of few-layer graphene with AB stacking *Phys. Rev.* B **73** 144427

[19] McCann E 2006 Asymmetry gap in the electronic band structure of bilayer graphene *Phys. Rev.* B **74** 161403

[20] Kumar S B and Guo J 2011 Multilayer graphene under vertical electric field *App. Phy. Lett* **98** 222101

[21] Avetisyan A A, Partoens B and Peeters F M 2009 Electric field tuning of the band gap in graphene multilayers *Phys. Rev.* B **79** 035421

[22] Castro E V, Novoselov K S, Morozov S V, Peres N M R, Lopes dos Santos J M B and Nilsson J *et al* 2007 Biased bilayer graphene: Semiconductor with a gap tunable by the electric field effect *Phys. Rev. Lett.* **99** 216802

[23] Yankowitz M, Wang F L, Lau C N and LeRoy B J 2013 Local spectroscopy of the electrically tunable band gap in trilayer graphene *Phys. Rev.* B **87** 165102

[24] Zhang W, Lin C T, Liu K K, Tite T, Su C Y and Chang C H *et al* 2011 Opening electrical band gap of bilayer graphene with molecular doping *ACS Nano* **5** 7517–24

[25] Ho Y H, Chiu Y H, Lin D H, Chang C P and Lin M F 2010 Magneto-optical selection rules in bilayer bernal graphene *ACS Nano* **4** 1465–72

[26] Abergel D S L and Fal'ko V I 2007 Optical and magneto-optical far-infrared properties of bilayer graphene *Phys. Rev.* B **75** 155430

[27] Charlier J C, Gonze X and Michenaud J P 1994 First-principles study of the stacking effect on the electronic-properties of graphite(s) *Carbon* **32** 289–99

[28] Charlier J C, Michenaud J P and Ph Lambin 1992 Tight-binding density of electronic states of pregraphitic carbon *Phys. Rev.* B **46** 4540–3

[29] Lu C L, Chang C P, Huang Y C, Lu J M, Hwang C C and Lin M F 2006 Low-energy electronic properties of the AB-stacked few-layer graphites *J. Phys. Condens. Matter* **18** 5849–59

[30] Craciun M F, Russo S, Yamamoto M, Oostinga J B, Morpurgo A F and Tarucha S 2009 Trilayer graphene is a semimetal with a gate-tunable band overlap *Nat. Nanotechnol.* **4** 383–8

[31] Cherkez V, Trambly de Laissardiere G, Mallet P and Veuillen J Y 2015 Van Hove singularities in doped twisted graphene bilayers studied by scanning tunneling spectroscopy *Phys. Rev.* B **91** 155428

[32] Lee Y, Velasco J, Tran D, Zhang F, Bao W and Jing L *et al* 2013 Broken symmetry quantum Hall states in dual-gated ABA trilayer graphene *Nano Lett.* **13** 1627–31

[33] Henriksen E A, Nandi D and Eisenstein J P 2012 Quantum Hall effect and semimetallic behavior of dual-gated ABA-stacked trilayer graphene *Phys. Rev.* X **2** 011004

[34] Cadden-Zimansky P, Jiang Z, Li Z Q, Tung L C, Schwartz M E and Takita M *et al* 2010 Interaction-induced shift of the cyclotron resonance of graphene using infrared spectroscopy *Phys. Rev. Lett.* **104** 067404

[35] Jiang Z, Henriksen E A, Tung L C, Wang Y J, Schwartz M E and Han M Y *et al* 2007 Infrared spectroscopy of Landau levels of graphene *Phys. Rev. Lett.* **98** 197403

[36] Plochocka P, Faugeras C, Orlita M, Sadowski M L, Martinez G and Potemski M *et al* 2008 High-energy limit of massless Dirac fermions in multilayer graphene using magneto-optical transmission spectroscopy *Phys. Rev. Lett.* **100** 087401

[37] Faugeras C, Amado M, Kossacki P, Orlita M, Kuhne M and Nicolet A A L *et al* 2011 Magneto-Raman scattering of graphene on graphite: electronic and phonon excitations *Phys. Rev. Lett.* **107** 036807

[38] Zhao H M, Lin Y C, Yeh C H, Tian H, Chen Y C and Xie D *et al* 2014 Growth and Raman spectra of single-crystal trilayer graphene with different stacking orientations *ACS Nano* **8** 10766–73

[39] Cong C X, Jung J, Cao B C, Qiu C Y, Shen X N and Ferreira A *et al* 2015 Magnetic oscillation of optical phonon in ABA- and ABC-stacked trilayer graphene *Phys. Rev.* B **91** 235403

[40] Tung L C, Cadden-Zimansky P, Qi J, Jiang Z and Smirnov D 2011 Measurement of graphite tight-binding parameters using high-field magnetoreflectance *Phys. Rev.* B **84** 153405

[41] Plochocka P, Solane P Y, Nicholas R J, Schneider J M, Piot B A and Maude D K *et al* 2012 Origin of electron-hole asymmetry in graphite and graphene *Phys. Rev.* B **85** 245410

[42] Zhang F, Sahu B, Min H and MacDonald A H 2010 Band structure of ABC-stacked graphene trilayers *Phys. Rev.* B **82** 035409

[43] Gusynin V P, Sharapov S G and Carbotte J P 2007 Anomalous absorption line in the magneto-optical response of graphene *Phys. Rev. Lett.* **98** 157402

[44] Koshino M and Ando T 2008 Magneto-optical properties of multilayer graphene *Phys. Rev.* B **77** 115313

[45] Ferrari A C, Meyer J C, Scardaci V, Casiraghi C, Lazzeri M and Mauri F *et al* 2006 Raman spectrum of graphene and graphene layers *Phys. Rev. Lett.* **97** 187401

IOP Publishing

Optical Properties of Graphene in Magnetic and Electric Fields

Chiun-Yan Lin, Thi-Nga Do, Yao-Kung Huang and Ming-Fa Lin

Chapter 5

ABC-stacked graphene

ABC stacking is a stable configuration in natural graphite [1, 2]. Few-layer ABC-stacked graphene is semimetallic with two partially flat bands slightly overlapping near the Fermi level [3–5]. The prominent peak of DOS near the Fermi level, coming from the partially flat bands, plays an important role in the low-frequency optical spectrum. In particular, energy bands exhibit sombrero-shaped dispersions near the nearest vertical interlayer hopping energy [6–9]. The corresponding absorption structures have been observed in the optical spectrum [10]. Furthermore, the middle-frequency spectrum is relevant to the band structure around the M point in the vicinity where the energy dispersions are significantly modified by interlayer atomic interactions. Under a magnetic field, it is intriguing that the LLs are inversely sequenced as a function of the field strength within the sombrero-shaped band structure, leading to frequent intragroup LL anticrossings accompanied by significant hybridizations of LLs. [11–14]. In general, the inter-LL transitions satisfy the particular selection rule $\Delta = \pm 1$ if the wavefunctions are well-behaved. However, extra optical selection rules are induced for the LLs with quantum modes in modulo 3, as a result of the LL hybridization. What is especially needed is to focus on the excitations within the LL anticrossing spectrum, where the diversified quantum modes lead to considerable peaks whose intensities are associated with the degree of hybridizations among the LLs.

5.1 Electronic and optical properties without external fields

ABC-stacked N-layer graphene consists of $2N$ carbon atoms in a primitive unit cell, see figure 2.1(e). Each of the graphene sheets is shifted by a distance of b' with respect to the lower (upper) neighboring layer along $+(-)\hat{x}$. The sublattice A of one layer is situated directly above the center of a hexagon in the adjacent lower layer, whereas the sublattice B lies above the lower-layer sublattice A. The intralayer and the interlayer atomic interactions in the tight-binding model are illustrated as β_i s, where $i = 0,1,2,3,4$ and 5. β_0 represents the nearest-neighbor intralayer hopping integral, β_1, β_3 and β_4

doi:10.1088/978-0-7503-1566-1ch5

© IOP Publishing Ltd 2017

correspond to adjacent interlayers, β_2 and β_5 are for the next-nearest interlayers. β_1 and β_2 couple two vertical sites and β_3, β_4 and β_5 couple non-vertical sites. Their values are listed as follows: $\beta_0 = -3.16$ eV, $\beta_1 = 0.36$ eV, $\beta_2 = -0.01$ eV, $\beta_3 = 0.32$ eV, $\beta_4 = 0.03$ eV, and $\beta_5 = 0.0065$ eV[56] [15].

The low-energy electronic properties of ABC-stacked trilayer graphene are characterized by three pairs of conduction and valence subbands in figure 5.1(a). Partially flat, sombrero-shaped and linear dispersions are displayed and respectively labeled $S_1^{c,v}$ (black), $S_2^{c,v}$ (red) and $S_3^{c,v}$ (blue). Being weakly dispersive in the vicinity of $E_F = 0$, the subbands of the first pair, $S_1^{c,v}$, consist of surface-localized states which are anisotropic around the K point as a result of trigonal warping [3–5]. The small energy difference between the two localized subbands is approximated as $2\beta_2 = 0.04$ eV. Moreover, the tight-binding calculations show that such a

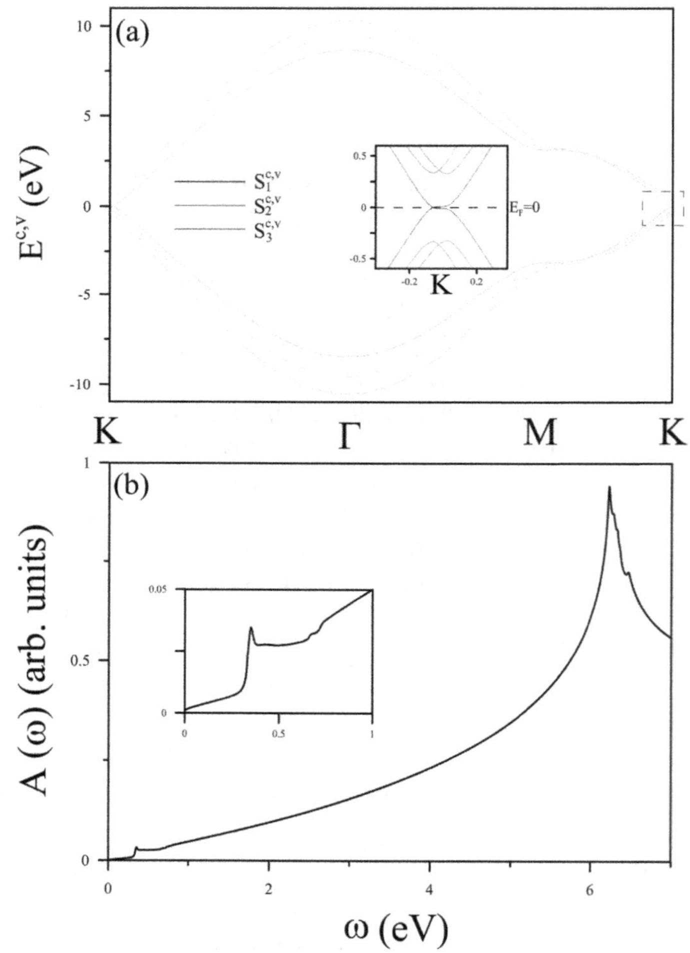

Figure 5.1. (a) The zero-field band structure and (b) the absorption spectrum of ABC-stacked trilayer graphene. In (a), the three subbands are labeled as $S_1^{c,v}$ (black), $S_2^{c,v}$ (red) and $S_3^{c,v}$ (blue). The insets of (a) and (b) show the zoomed-in view at low energies.

characteristic of the surface localized states disappears as the wave vector moves away from the K point. At $E_F = 0$, the three crossing points in a circle are identified as Dirac points [16]: one in the KΓ line and another two whose rotation angles are relative to the former by 120° and −120°. Away from $E_F = 0$, the sombrero-shaped and linear subbands cross each other near the K point at energies $\sim\pm\beta_1$ ($=\pm0.36$ eV), where the crossing point corresponds to a local minimum of the former and a local maximum of the latter. In particular, within the bottom (~0.34 eV) and tip (~0.36 eV) of the sombrero-shaped subband, two closed circular loops at a specific energy lead to a doubly degenerate state with different Landau orbitals under a magnetic field. This would cause unusual characteristics for certain excitation channels. Interlayer atomic interactions also modify and diversify the band structures around β_0. This is mainly caused by β_3 and β_5, in which some saddle-point states are distributed away from the M point over a small energy range and some are shifted along the ΓM direction. Recently, the theoretical band structure was experimentally confirmed by STS and ARPES [17–19].

The optical absorption spectrum correlates with the particular energy spectrum in both regions of $\omega < 2\beta_1$ around the K point, and of $\omega \sim 2\beta_0$ around the M point according to equation (2.3), as shown in figure 5.1(b). The low-frequency spectrum is dominated by excitations from the former region, which are mainly affected by the interlayer atomic interactions β_1 and β_2. There is one obvious asymmetric peak at $\omega \simeq 0.35$ eV, one observable peak at $\omega \simeq 0.68$ eV and one shoulder at $\omega \simeq 0.72$ eV, respectively, due to the vertical transitions between the surface-localized and sombrero-shaped subbands, between the valence and conduction sombrero-shaped subbands, and between two linear subbands. Specifically, the latter two structures, with a small frequency difference, might merge together in optical measurements. On the other hand, 3 × 3 kinds of excitation channels from the saddle-points in the latter region induce several obvious peaks and shoulder structures for the middle-frequency spectrum $\sim 2\beta_0$, in which the interlayer atomic interactions β_3 and β_5 play an important role. A prominent symmetric peak appears at $\omega \simeq 6.21$ eV, as a result of transitions between the saddle points of the subbands $S_1^{c,v}$. The two peaks at $\omega \simeq 6.28$ eV and $\omega \simeq 6.46$ eV are responsible for the saddle-point transitions of pairs of subbands $S_1^{c,v}$ and $S_3^{c,v}$. Moreover, the characteristic peak at $\omega \simeq 6.32$ eV is associated with the saddle-point of S_2^v [10]. Recently, infrared reflection spectroscopy and absorption spectroscopy verified the low-frequency optical properties, showing clear evidence of two characteristic peaks associated with the partially flat and sombrero-shaped subbands [11]. Based on the specific infrared conductivities, infrared scattering scanning near-field optical microscopy can distinguish the ABC stacking domains with nanoscale resolution from the others [11]. Furthermore, the whole frequency spectrum deserves closer examination to verify the band structure over a wide energy region, and the measured features are useful for identifying the stacking configuration [11].

5.2 Magneto-electronic and magneto-optical properties

The $4NR_B$ Peierls tight-binding functions based on the periodical atoms in the rectangular unit cell are expressed as: $\{|A_{1\mathbf{k}}^1\rangle, |B_{1\mathbf{k}}^1\rangle, |A_{1\mathbf{k}}^2\rangle, |B_{1\mathbf{k}}^2\rangle \cdots |A_{1\mathbf{k}}^N\rangle, |B_{1\mathbf{k}}^N\rangle \cdots |A_{2R_B\mathbf{k}}^N\rangle, |B_{2R_B1\mathbf{k}}^N\rangle\}$.

The non-zero elements of the Peierls tight-binding Hamiltonian matrix are given by

$$\langle B^L_{m\mathbf{k}}|H|A^{L'}_{m'\mathbf{k}}\rangle$$
$$=\beta_0(t_{1,k}\delta_{m,m'}+t_{4,k}\delta_{m,m'-1})\delta_{L,L'}+\beta_3(t^*_{3,k}\delta_{m,m'+1}+t^*_{4,k}\delta_{m,m'})\delta_{L,L'-1}\ \text{for}\ L=3l-2,$$
$$\langle B^L_{m\mathbf{k}}|H|A^{L'}_{m'\mathbf{k}}\rangle$$
$$=\beta_0(t_{3,k}\delta_{m,m'+1}+t_{4,k}\delta_{m,m'})\delta_{L,L'}+\beta_3(t^*_{2,k}\delta_{m,m'}+t^*_{4,k}\delta_{m,m'-1})\delta_{L,L'-1}\ \text{for}\ L=3l-1,$$
$$\langle B^L_{m\mathbf{k}}|H|A^{L'}_{m'\mathbf{k}}\rangle$$
$$=\beta_0(t_{2,k}\delta_{m,m'}+t_{4,k}\delta_{m,m'-1})\delta_{L,L'}+\beta_3(t^*_{1,k}\delta_{m,m'}+t^*_{4,k}\delta_{m,m'-1})\delta_{L,L'-1}\ \text{for}\ L=3l,$$
$$\langle B^L_{m\mathbf{k}}|H|A^{L'}_{m'\mathbf{k}}\rangle=\beta_2\delta_{m,m'}\delta_{L,L'-2}+\beta_1\delta_{m,m'}\delta_{L,L'-1}\ \text{for any L},$$
$$\langle B^L_{m\mathbf{k}}|H|B^{L'}_{m'\mathbf{k}}\rangle=\langle A^{L+2}_{m\mathbf{k}}|H|A^{L'+2}_{m'\mathbf{k}}\rangle=\beta_4(t_{1,k}\delta_{m,m'}+t_{4,k}\delta_{m,m'-1})\delta_{L,L'-1}\ \text{for}\ L=3l-2,$$
$$\langle A^L_{m\mathbf{k}}|H|A^{L'}_{m'\mathbf{k}}\rangle=\langle B^{L+1}_{m\mathbf{k}}|H|B^{L'+1}_{m'\mathbf{k}}\rangle=\beta_4(t_{3,k}\delta_{m,m'+1}+t_{4,k}\delta_{m,m'})\delta_{L,L'-1}\ \text{for}\ L=3l-2,$$
$$\langle A^L_{m\mathbf{k}}|H|B^{L'}_{m'\mathbf{k}}\rangle=\langle B^{L'+2}_{m\mathbf{k}}|H|B^{L+2}_{m'\mathbf{k}}\rangle=\langle A^{L'+1}_{m\mathbf{k}}|H|A^{L+1}_{m'\mathbf{k}}\rangle$$
$$=\beta_5(t_{3,k}\delta_{m,m'+1}+t_{4,k}\delta_{m,m'})\delta_{L,L'-2}\ \text{for L}=3l-2,$$
$$\langle A^L_{m\mathbf{k}}|H|A^{L'}_{m'\mathbf{k}}\rangle=\langle B^{L+1}_{m\mathbf{k}}|H|B^{L'+1}_{m'\mathbf{k}}\rangle=\beta_4(t_{2,k}\delta_{m,m'}+t_{4,k}\delta_{m,m'-1})\delta_{L,L'-1}\ \text{for L}=3l-1,$$
$$\langle B^L_{m\mathbf{k}}|H|B^{L'}_{m'\mathbf{k}}\rangle=\langle A^{L+2}_{m\mathbf{k}}|H|A^{L'+2}_{m'\mathbf{k}}\rangle=\langle A^{L'+1}_{m\mathbf{k}}|H|B^{L+1}_{m\mathbf{k}}\rangle$$
$$=\beta_5(t_{2,k}\delta_{m,m'}+t_{4,k}\delta_{m,m'-1})\delta_{L,L'+2}\ \text{for L}=3l,$$
$$\langle A^L_{m\mathbf{k}}|H|A^{L'}_{m'\mathbf{k}}\rangle=\langle B^{L+1}_{m\mathbf{k}}|H|B^{L'+1}_{m'\mathbf{k}}\rangle=\langle A^{L'+2}_{m\mathbf{k}}|H|B^{L+2}_{m'\mathbf{k}}\rangle$$
$$=\beta_5(t_{1,k}\delta_{m,m'}+t_{4,k}\delta_{m,m'-1})\delta_{L,L'+2}\ \text{for L}=3l.$$

The magnetically quantized LLs of ABC-stacked trilayer graphene can be divided into three groups according to three zero-field subbands and the spatial characteristics of the wavefunctions, as indicated by black, red and blue in figure 5.2 at $B_0 = 40$ T. The valence and conduction LLs are asymmetric with respect to $E_F = 0$ due to interlayer atomic interactions. Each level is four-fold degenerate under the inversion symmetry. The A^l_o subenvelope functions of the 1/6- (4/6-) localized states ($l = 1, 2; 3$) are out of phase with the B^l_o subenvelope functions of the 2/6- (5/6-) localized states. The wavefunctions at the 2/6 unit cell are illustrated. As a result of being quantized by the partially flat subband, the first group of LLs are initiated from $E_F = 0$, in the vicinity of which the first three LLs, coming from the Dirac points, are slightly separated by the interlayer hoppings, while the other higher-order LLs are distributed away from $E_F = 0$ [14, 16, 20]. The dominating sublattice in the first group corresponds to the outermost B^1 sites of the surface-localized states. Similarly, the corresponding outermost A^3 sites can be used to define the quantum numbers of the other degenerate states localized at the 1/6 and 4/6 unit cell. The quantum numbers of LLs are counted in ascending order from $E_F = 0$, as evidenced by the dominating subenvelope function B^1_o. Meanwhile, resulting from the sombrero-shaped and linear subbands, the LLs in the second and third groups are initiated by the intersection energy $\sim\beta_1$ of the subbands, and

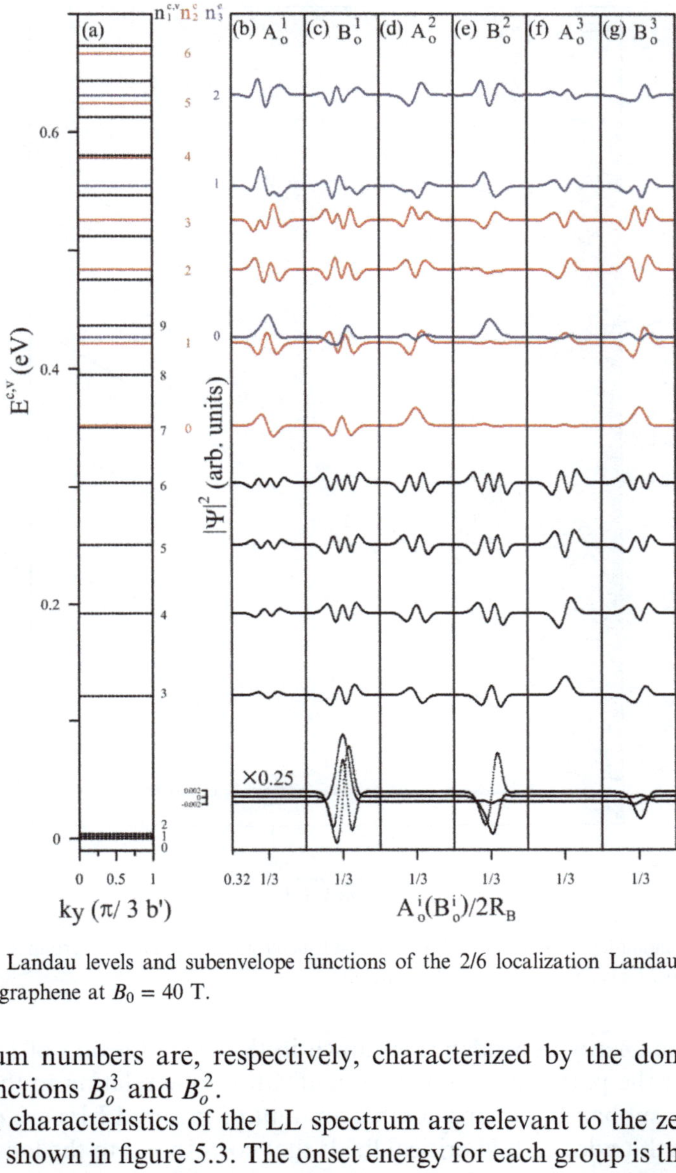

Figure 5.2. The Landau levels and subenvelope functions of the 2/6 localization Landau states for ABC-stacked trilayer graphene at $B_0 = 40$ T.

their quantum numbers are, respectively, characterized by the dominating sub-envelope functions B_o^3 and B_o^2.

The main characteristics of the LL spectrum are relevant to the zero-field band structure, as shown in figure 5.3. The onset energy for each group is the same as the band state energy of the respective subband near the K point. The level spacings are determined by the DOS, that is to say, the LLs are densely packed with the increment of the DOS. Accordingly, the intensive formation of the LLs near $\pm\beta_1$ is attributed to the crossover of the subbands in figure 5.1. The first and third groups present the monotonically B_0- dependent energies. Near $E_F = 0$, the three LLs that are insensitive to the field strength reflect the characteristics of the Dirac points. They are confined in a small energy spacing between the two partially flat subbands, with the separated energies mainly determined by the specific interlayer atomic interaction γ_4. This sequence of the first group of LLs was confirmed by the measurements of the Hall conductance [21]. However, a complicated pattern of

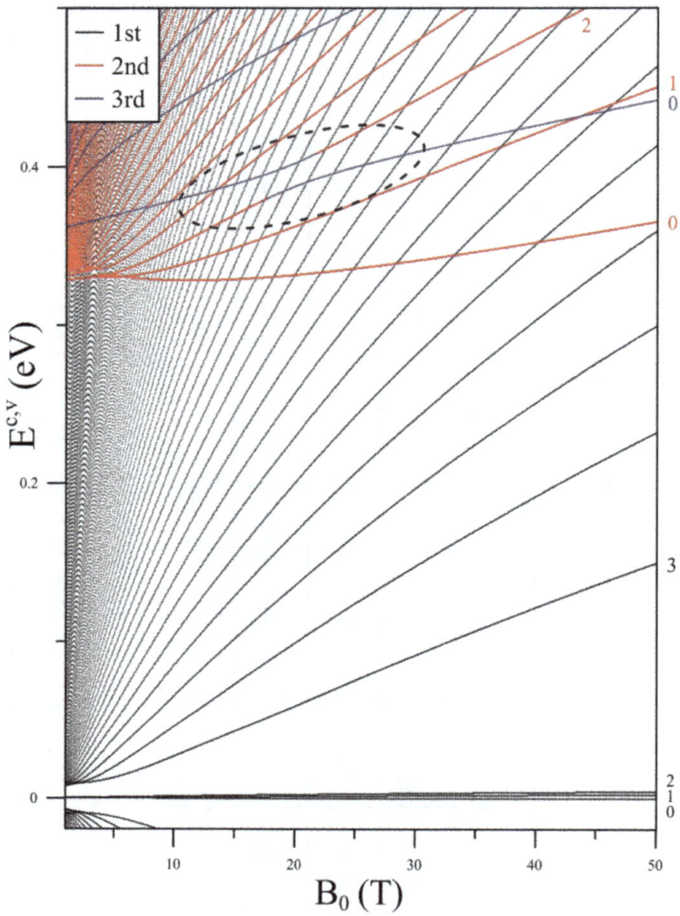

Figure 5.3. The Landau-level spectrum of ABC-stacked trilayer graphene. The intergroup anticrossing of the $n_2^c = 2$ and $n_3^c = 0$ LLs is shown in the dashed circle.

non-monotonic energy dependence is shown in the second group of LLs, which is responsible for the particular magnetic quantization rule relating to the sombrero-shaped energy subbands. A considerable amount of inverted LLs are observed for the field strength under certain critical fields that correspond to the local minima in the spectrum.

The relationship of the inverted LLs and the critical magnetic field is dominated by the interplay between the magnetic degeneracy and the total states enclosed at the sombrero surface, as shown in figure 5.4(a). Two kinds of B_0-dependent energies correspond to two kinds of degenerate Landau orbitals, i.e. the proportional and the inversely proportional dependences, for the respective outer and inner circular constant energy loops on the surface of the sombrero-shaped subbands. A weak magnetic field aggregates the electronic states near the vertex of the sombrero-shaped subband. Therefore, as more states on the inner surface are magnetically quantized, the LLs display inversely proportional energy dependence on the field

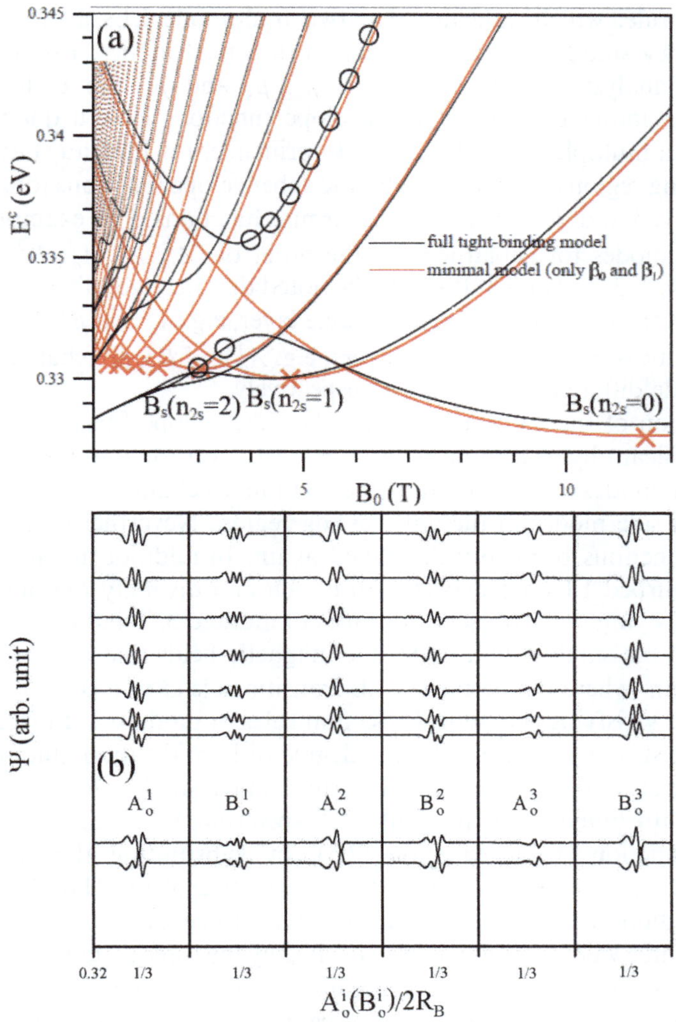

Figure 5.4. The Landau levels of the second group of ABC-stacked trilayer graphene based on the minimum model (red) and the full tight-binding model (black). Red cross symbols indicate the critical fields $B_s(n_{2s})$ for specific Landau levels. (b) The evolution of the subenvelope functions for the anticrossing Landau states marked in (a).

strength. In contrast, the magnetic quantification associated with the outer surface is introduced to the B_0-proportional LLs. It can be deduced that at a critical field all the inner-surface states are exactly quantized to the inverted LLs, and the inverse of the critical field strength exhibits a periodicity with the total numbers of inverted LLs because of the constant enclosed states. Apparently, the particular band structures characterize the magneto-absorption spectrum.

As a signature of the ABC-stacked graphene, the magnetic-field-dependent spectrum exhibits many anticrossings of the second group of LLs, as shown in figure 5.4(a). The reason for this is attributed to the Wigner–von Neumann non-crossing rule, which indicates that two multimode LLs avoid crossing each other once they simultaneously have certain identical modes. From a qualitative perturbation analysis, the three factors, β_2, β_3 and β_5, are deduced to cause hybridization among LLs. Their subenvelope functions have a quantum number difference by a multiple of 3 [14, 22–24]. In principle, for Landau states away from the anticrossing regions in Figure 5.4(b), the subenvelope functions for all sublattices are well-behaved and described by the Hermite functions. For example, of $n_2^c = 3$, the quantum modes for sublattices on the order of $\{A_o^1, B_o^1, A_o^2, B_o^2, A_o^3, B_o^3\}$ are, respectively, 4, 5, 3, 4, 2 and 3. It should be noted that the two subenvelope functions established on the two sublattices in the same layer, e.g. A_o^1 and B_o^1, and those on the vertical sublattices between adjacent layers, e.g. A_o^1 and B_o^2, are characterized by the quantum-mode differences of 1 and 0, respectively.

At the region of the LL anticrossings, the subenvelope functions are perturbed due to a significant hybridization with $n_2^c = 0$ LL, according to a difference of 3 in their quantum modes, as evidenced by the continuous change for the components of the main and side modes in the anticrossing region. Nevertheless, away from this region, the functions become well-defined again. In addition to the well-behaved LLs, the perturbed LLs are expected to enrich and diversify the magneto-optical properties satisfying the extra selection rules of modulo 3. What is especially needed is to focus on the small region under the magnetic fields that are weaker than the critical strength, where the perturbed LL transitions lead to a considerable amount of extra peaks satisfying the optical selection rules of modulo 3. Furthermore, their intensities are strongly increased by the degree of hybridization among the LLs.

A feature-rich optical absorption spectrum is presented in ABC-stacked trilayer graphene, as attributed to the particular LL spectrum. One can obtain the spectral absorption function $A(\omega)$, equation (2.3), by substituting the velocity matrix evaluated from the wavefunctions and the derivative of the Hamiltonian over all possible transitions in the first Brillouin zone. It is shown that occupied Landau states in any of the three valence groups can be excited by the three conduction groups, once the excitation channels satisfy the generalized selection rules (see chapter 2). The absorption peaks can be classified into nine categories resulting from three intragroup and six intergroup LL transitions, as shown in figure 5.5(a) for $B_0 = 40$ T, where the black, red and green numbers, respectively, indicate the excitations from the first, second and third LL groups. In general, many of the absorption peaks exhibit twin-peak structures as a result of the two kinds of excitation channels with a tiny energy difference in the asymmetric LL spectrum. However, for some categories, the threshold channels that come exclusively from the three Dirac-point-related LLs, $n_1^v = 0$, $n_1^{c,v} = 1$ and $n_1^c = 2$, give rise to single-peak structures according to the Fermi–Dirac distribution, i.e. $\omega_{1,1}^{3^v2^c}$, $\omega_{1,3}^{0^v0^c}$ and $\omega_{2,1}^{1^v2^c}$. The frequency spacing of the twin peaks may be observable, providing a way to verify the asymmetry between the valence and conduction LLs. It should be noted that the magneto-absorption spectrum exhibits some unusual absorption peaks due to the transitions between the perturbed LLs in the region of the LL anticrossings, as indicated by the dashed ellipse in figure 5.3. Induced by the side modes of the hybridized LL ($n_3^v = 0$), such

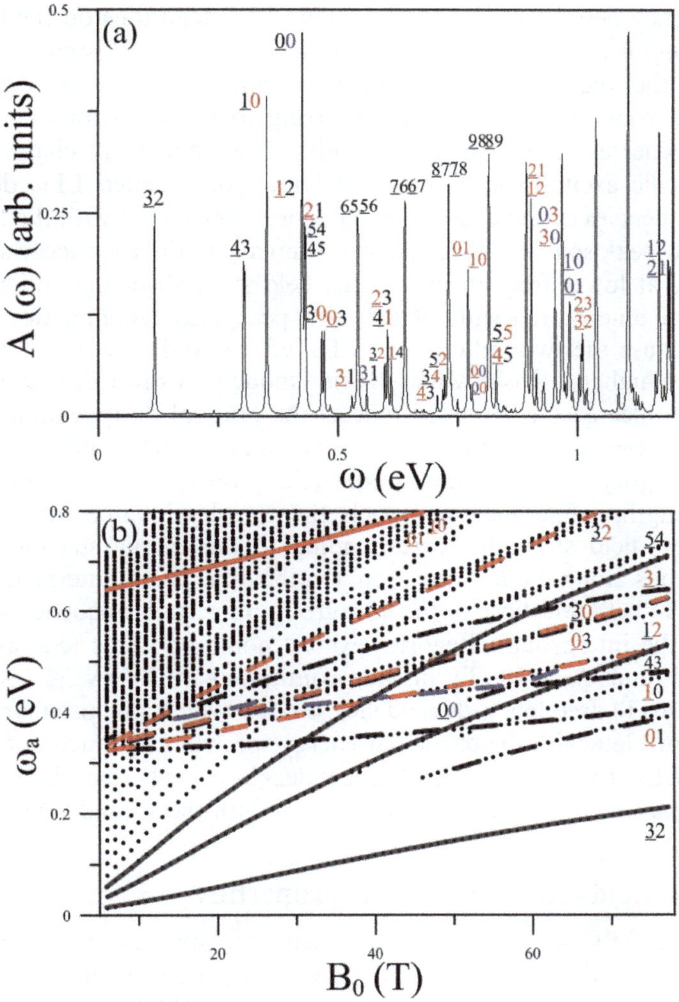

Figure 5.5. (a) The magneto-absorption spectrum of ABC-stacked trilayer graphene under B_0 = 40 T. The black, red and green numbers, respectively, indicate the excitations from the first, second and third LL groups. (b) The B_0-dependent magneto-absorption frequencies. The solid and dashed curves indicate the excitation channels of the intragroup and intergroup categories, respectively.

extra peaks satisfy the special selection rules of modulo 3. Moreover, the induced peak intensity varies with the degree of the hybridization. For example, considering the transitions from the $n_3^v = 0$ LL, one finds an extra twin peak near 0.96 eV that includes the $n_3^v = 0 \rightarrow n_2^c = 3$ and $n_2^v = 3 \rightarrow n_3^c = 0$ excitations, in addition to the regular twin peak $\backsim$0.78 eV resulting from the $n_3^v = 0 \rightarrow n_2^c = 0$ and $n_2^v = 0 \rightarrow n_3^c = 0$ excitations. On the other hand, the intensities of the regular absorption peaks in ABC-stacked graphene are irregular with respect to the photon frequency, in contrast to the uniformity of monolayer graphene. The main reason for this is attributed to the relationship of the phase and amplitude between the subenvelope functions, depending strongly on the stacking configurations.

The frequency dependence of the low-lying LL excitations on the field strength deserves a closer examination to explore the spectral characteristics of ABC-stacked graphene. Neither the intergroup nor the intragroup categories can be described as a simple square root or linear formula referring to the behavior of monolayer or bilayer graphene, as shown in figure 5.5(b). The single-peak channels, resulting from the specific excitations of the three Dirac-point-related LLs, determine the threshold frequencies of the absorption categories associated with the first group of LLs. The twin-peak spacings, due to the asymmetry in the LL energy spectrum, are more obvious at lower frequencies. At the field strength of the anticrossing point ($B_0 \simeq 20$ T), a blue or red shift of the twin peak also accompanies the intensity variation, because the two anticrossing LLs, $n_2^c = 0$ and $n_3^c = 1$, are reversed. The extra peak has higher intensity when the side mode gets stronger with an increment of B_0, and the decreasing intensity reflects the gradual disappearance of the side mode after B_0 passes through the strongest anticrossing point [25]. In the first intragroup category, the absorption frequencies grow quickly with an increment of the field strength, while they approach the threshold value of $\omega_{32} \simeq 2\beta_2$ at a sufficiently low field strength, corresponding to the energy spacing between the partial flat bands along the $K\Gamma$ direction. On the other hand, intergroup absorption peaks appear in between those of the intragroup. In certain frequency and magnetic field ranges, the intergroup categories exhibit non-monotonic (e.g. $\omega_{21}^{1^v 0^v}$; 0.30 eV $\lesssim \omega \lesssim$ 0.35 eV and $B_0 \simeq 20$ T) or non-continuous (0.30 eV $\lesssim \omega \lesssim$ 0.5 eV and $B_0 \simeq 20$, 55; 70 T) frequency dependence on B_0, which are peculiar phenomena that closely correlate with the particular energy spectrum that includes inverted and anticrossing LLs. Furthermore, both dependences can also be observed for those higher frequency excitation channels associated with the second group of LLs.

5.3 Electric-field-enriched optical properties

Significantly, in ABC-stacked trilayer graphene, the application of a perpendicular electric field leads to drastic changes in the energy dispersion, band gap, band spacing, anticrossing and band-edge states shown in figure 5.6(a) [26], as well as enriching the absorption spectra with different peak structures in figure 5.6(b). As a result of the special surface-localized states near the Fermi level, the energy gap is proportional to the potential difference between two outermost layers under a moderate electric field. The first absorption peak directly reflects the optical gap transitions and is used to determine the value of E_g, as observed in AB-stacked bilayer graphene. The second peak, associated with the partially flat subband and the distorted sombrero-shaped subband, exhibits an observable blue shift. By contrast, the spectral characteristics of AA- and AB-stacked trilayer graphene remain similar under the electric field. In AB-stacked trilayer graphene, the **E** field increases the overlap between the valence and conduction bands, and enhances the optical conductivity; in AA-stacked trilayer graphene, it critically changes the Dirac-point energies, the Fermi momenta and Fermi velocity and modulates the forbidden optical transition regions.

The four-fold degenerate LLs are split into two-fold degenerate LLs because of the **E**-induced destruction of the mirror symmetry in the xy-plane and the

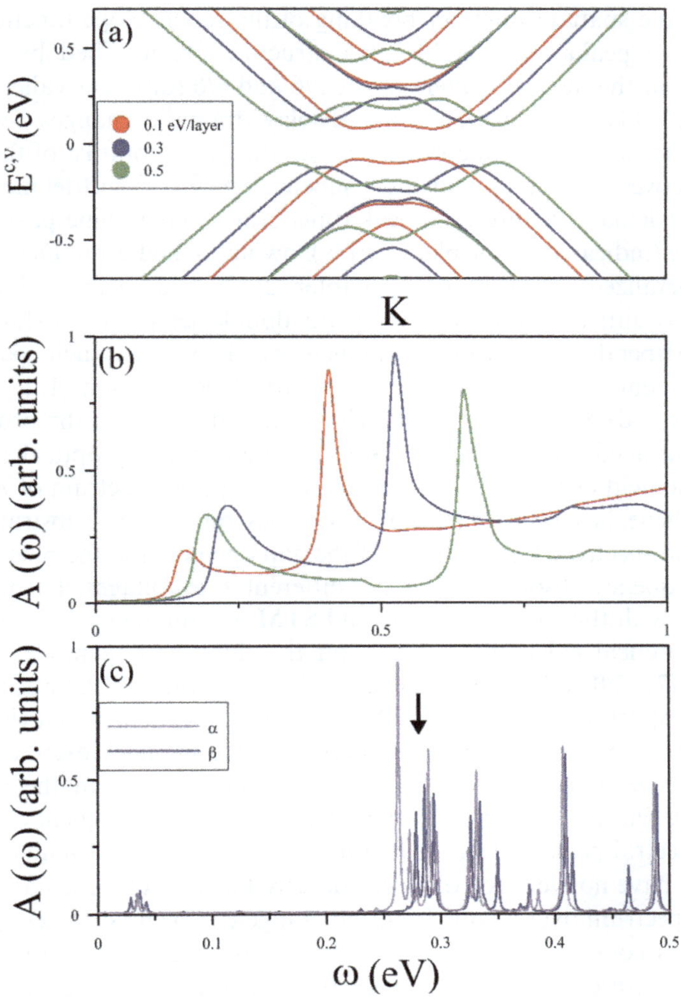

Figure 5.6. The E_0-enriched (a) band structure, (b) absorption spectrum and (c) magneto-absorption spectrum of ABC-stacked trilayer graphene.

non-equivalence of the two sublattices in identical projections. Each group of LLs becomes two subgroups in which the splitting LLs are characterized by two different localization centers of the 1/6 and 2/6 unit cell (4/6 and 5/6 unit cell). The dominating sublattice, in determining quantum numbers, remains unchanged for each subgroup, i.e. two inversely symmetric sublattices for two split subgroups. As more inter-LL channels are excited, e.g. $B_0 = 40$ T and $E_0 = 0.3$ eV/layer, the absorption spectrum shows double peaks, identified as α and β types for the inter-LL transitions from the former and latter subgroups, respectively (the blue and purple curves in figure 5.6(c)). At a moderate applied electric field ~0.06 eV/layer, the coexistence of the twin- and double-peak structures is displayed, as shown in figure 3.2 of [13]. However, with a further increment of V_g, a lot of prominent twin-peak absorption peaks gradually change into double-peaks, which is a transition that is attributed to the spatial symmetry breaking of the subenvelope functions. In other words, the two peaks in a double-peak structure are ascribed by the inter-LL transitions from the distinct centers of the 2/6 and 4/6 (different valleys).

The α- and β-type absorption peaks display dramatic changes because of the breaking of the stacking symmetry that causes the sharp contrast of the amplitudes and phases between their non-equivalent subenvelope functions (detailed discussions in [13]). In particular, the threshold peak transforms from a single-peak to a double-peak structure (indicated by the black arrow), owing to the Fermi–Dirac distribution and the generalized selection rule (chapter 2). The electric field also induces intragroup LL anticrossings and thus extra double peaks that exhibit abnormal electric-field-dependent absorption frequencies. However, such peaks can be suppressed or enhanced by varying the electric field strength [13]. The induced peak intensity is determined by the weight of the side mode in the progressive LL. The aforementioned optical characteristics can be verified by optical spectroscopy.

The electric-field-dependent low-frequency absorption spectrum of ABC-stacked trilayer graphene has been observed in samples on SiO_2/Si substrates [13]. The results provide spectroscopic evidence of the gap opening and the blue (or red) shift of the low-frequency absorption peaks at different gate voltages. The measured gap is compatible with that from transport and STM measurements [17] and consistent with the theoretical calculations based on the tight-binding model over a wide energy range [27–29]. While the sequence of the first group of LLs was confirmed by the Hall transport measurements [30–32], further experimental evidence of the magneto-optical spectrum is needed for verifying the inter-LL excitations in ABC-stacked graphene, which display different characteristics from the massless and massive Dirac fermions in terms of the intensity, number, frequency and selection rule of the spectral peaks. Also, the features due to electric-field-induced symmetry breaking that have not yet been observed deserve further experimental examination. The optical spectrum over a wide frequency range can be used to verify the total of nine categories consisting of single and twin peaks, i.e. three intragroup and six intergroup categories. The peak intensities are different, and in general, the intragroup ones are higher than the intergroup ones. In addition, as a result of the LL anticrossing, the extra peaks exhibit a small range of non-continuous intensities and frequencies during the variation of the magnetic field (figure 5.5(b)), in which the strongest intensity and blue and red shifts occur at the anticrossing point. At a sufficiently low magnetic field, the characteristic peaks with non-monotonic frequencies reflect the magnetic quantization of the sombrero-shaped subbands, and the first single channel threshold can be used to determine the vertical next-nearest interlayer atomic interaction β_2.

References

[1] Bernal J D 1924 The structure of graphite *Proc. R. Soc. London, Ser.* A **106** 749

[2] Lipson H and Stokes A R 1942 The structure of graphite *Proc. R. Sot. London, Ser.* A **181** 101

[3] Guinea F, Castro Neto A H and Peres N M R 2006 Electronic states and Landau levels in graphene stacks *Phys. Rev.* B **73** 245426

[4] Kopnin N B, Heikkila T T and Volovik G E 2011 High-temperature surface superconductivity in topological flat-band systems *Phys. Rev.* B **83** 220503

[5] Wang H, Gao J H and Zhang F C 2013 Flat band electrons and interactions in rhombohedral trilayer graphene *Phys. Rev.* B **87** 155116

[6] Castro Neto A H, Guinea F, Peres N M R, Novoselov K S and Geim A K 2009 The electronic properties of graphene *Rev. Mod. Phys.* **81** 109–62

[7] Latil S and Henrard L 2006 Charge carriers in few-layer graphene films *Phys. Rev. Lett.* **97** 036803

[8] Xiao R, Tasnadi F, Koepernik K, Venderbos J W F, Richter M and Taut M 2011 Density functional investigation of rhombohedral stacks of graphene: topological surface states, nonlinear dielectric response, and bulk limit *Phys. Rev.* B **84** 165404

[9] Zhang F, Sahu B, Min H and MacDonald A H 2010 Band structure of ABC-stacked graphene trilayers *Phys. Rev.* B **82** 035409

[10] Lui C H, Li Z, Chen Z, Klimov P V, Brus L E and Heinz T F 2011 Imaging stacking order in few-layer graphene *Nano Lett.* **11** 164–9

[11] Mak K F, Shan J and Heinz T F 2010 Electronic structure of few-layer graphene: experimental demonstration of strong dependence on stacking sequence *Phys. Rev. Lett.* **104** 176404

[12] Lin Y P, Lin C Y, Ho Y H, Do T N and Lin M F 2015 Magneto-optical properties of ABC-stacked trilayer graphene *Phys. Chem. Chem. Phys.* **17** 15921–7

[13] Lin Y P, Lin C Y, Chang C P and Lin M F 2015 Electric-field-induced rich magneto-absorption spectra of ABC-stacked trilayer graphene *RSC Adv* **5** 80410

[14] Lin C Y, Wu J Y, Chiu Y H, Chang C P and Lin M F 2014 Stacking-dependent magnetoelectronic properties in multilayer graphene *Phy. Rev.* B **90** 205434

[15] Charlier J C, Michenaud J P and Lambin Ph 1992 Tight-binding density of electronic states of pregraphitic carbon *Phys. Rev.* B **46** 4540–3

[16] Koshino M and McCann E 2009 Trigonal warping and Berry's phase $N\pi$ in ABC-stacked multilayer graphene *Phys. Rev.* B **80** 165409

[17] Yankowitz M, Wang F L, Lau C N and LeRoy B J 2013 Local spectroscopy of the electrically tunable band gap in trilayer graphene *Phys. Rev.* B **87** 165102

[18] Bao W, Jing L, Velasco J, Lee Y, Liu G and Tran D *et al* 2011 Stacking-dependent band gap and quantum transport in trilayer graphene *Nat. Phys.* **7** 948–52

[19] Craciun M F, Russo S, Yamamoto M, Oostinga J B, Morpurgo A F and Tarucha S 2009 Trilayer graphene is a semimetal with a gate-tunable band overlap *Nat. Nanotechnol.* **4** 383–8

[20] Ho C H, Ho Y H, Chiu Y H, Chen Y N and Lin M F 2011 Magneto-electronic properties of rhombohedral trilayer graphene: Peierls tight-binding model *Ann. Phys.* **326** 721–39

[21] Henriksen E A, Nandi D and Eisenstein J P 2012 Quantum Hall effect and semimetallic behavior of dual-gated ABA-stacked trilayer graphene *Phys. Rev.* X **2** 011004

[22] Koshino M and McCann E 2011 Landau level spectra and the quantum Hall effect of multilayer graphene *Phys. Rev.* B **83** 165443

[23] Lin Y P, Wang J, Lu J M, Lin C Y and Lin M F 2014 Energy spectra of ABC-stacked trilayer graphene in magnetic and electric fields *RSC Adv.* **4** 56552–60

[24] Inoue M 1962 Landau levels and cyclotron resonance in graphite *J. Phys. Soc. Jpn* **17** 808

[25] Ho Y H, Chiu Y H, Lin D H, Chang C P and Lin M F 2010 Magneto-optical selection rules in bilayer Bernal graphene *ACS Nano.* **4** 1465–72

[26] Lui C H, Li Z, Mak K F, Cappelluti E and Heinz T F 2102 Observation of an electrically tunable band gap in trilayer graphene *Nat. Phys.* **7** 944–7

[27] Avetisyan A A, Partoens B and Peeters F M 2009 Electric field tuning of the band gap in graphene multilayers *Phys. Rev.* B **79** 035421

[28] Lu C L, Chang C P, Huang Y C, Lu J M, Hwang C C and Lin M F 2006 Low-energy electronic properties of the AB-stacked few-layer graphites *J. Phys.: Condens. Matter* **18** 5849–59

[29] McCann E 2006 Asymmetry gap in the electronic band structure of bilayer graphene *Phys. Rev.* B **74** 161403

[30] Kumar A, Escoffier W, Poumirol J M, Faugeras C, Arovas D P and Fogler M M *et al* 2011 Integer quantum Hall effect in trilayer graphene *Phys. Rev. Lett.* **107** 126806

[31] van Elferen H J, Veligura A, Tombros N, Kurganova E V, van Wees B J and Maan J C *et al* 2013 Fine structure of the lowest Landau level in suspended trilayer graphene *Phys. Rev.* B **88** 121302

[32] Zhang L, Zhang Y, Camacho J, Khodas M and Zaliznyak I 2011 The experimental observation of quantum Hall effect of $l = 3$ chiral quasiparticles in trilayer graphene *Nat. Phys.* **7** 953–7

Chapter 6

AAB-stacked graphene

AAB stacking, which is the second trilayer stacking characteristic, is one of the important stacking configurations. This system has been successfully synthesized by different experimental methods, such as electrostatic manipulation STM [1–3], mechanical cleavage [4, 5], the liquid phase exfoliation of natural graphite [6], and growth on a silicon carbide substrate [7] or metal surface [8]. Specifically, the stacking configuration can be continuously changed by the STM tip. The first principle calculations show that the ground state energies of the four stacking configurations are −55.832866 eV for AAA stacking, −55.857 749 eV for AAB stacking, −55.862 386 eV for ABA stacking and −55.864 039 eV for ABC stacking. There exist only small energy differences between the four typical types of stacking, and the AAB stacking is more stable than that of AAA. As for trilayer AAB stacking, only a few theoretical studies are focused on the low-energy electronic [5, 9] and optical properties [10]. Nine interlayer hopping integrals are required to characterize the unique band structure (section 6.1). This clearly illustrates that stacking configurations dominate the interlayer atomic interactions and thus the essential physical properties.

The electronic structure consists of three pairs of valence and conduction bands corresponding to oscillatory, sombrero-shaped and parabolic. For the oscillatory pair, each subband has three circular constant energy contours in the energy wave vector space. This was first revealed in AAB stacking, implying that it is exclusive to trilayer systems. A similar characteristic is present for sombrero-shaped bands with two constant energy loops. These unusual energy dispersions will induce complex magnetic quantization as a result of the multi-constant-energy loops. This will be reflected in the extraordinary features of the magneto-optical spectra. Without the external fields, any excitation channels between the valence and conduction bands are available; furthermore, the asymmetric and symmetric peaks are, respectively, shown in the low- and middle-frequency absorption spectrum. The magneto-absorption spectra are greatly enriched by three intragroup and six intergroup

doi:10.1088/978-0-7503-1566-1ch6

© IOP Publishing Ltd 2017

inter-LL optical excitations. The special relations in the phases and amplitudes of the LL wavefunctions result in a significant difference in intensity between the intra- and intergroup transition peaks. The absorption spectra exhibit single- or twin-peak structures determined by the quantum modes, LL energy spectra and fermion distribution. Moreover, the splitting LLs arising from the destruction of inversion symmetry can diversify the magneto-absorption spectra. These characteristics are absent in other stacking systems. In general, the excitations relating to LLs with well-behaved wavefunctions satisfy the general optical selection rule. Nevertheless, many extra single peaks are caused by sufficiently strong side modes during LL anticrossings. The important differences between the four types of typically stacked trilayer graphene mostly lie in the number of excitation categories and the main features of the absorption peaks.

6.1 Electronic and optical properties without external fields

For the AAB stacking, the first two layers are arranged as an AA-stacked configuration while the third one is the shift of the first (or the second) layer by a distance of b along the armchair direction. The zero-field Hamiltonian built from the six tight-binding functions of the $2p_z$ orbitals is dominated by the intralayer and the interlayer atomic interactions $\gamma_i' s$. In order to simulate the unusual energy dispersions, ten atomic hopping integrals are used in the calculations (Figure 2.1). $\gamma_0 = -2.569$ eV represents the nearest-neighbor intralayer atomic interaction, $\gamma_1 = -0.263$ eV and $\gamma_2 = 0.032$ eV are related to the interlayer atomic interactions between the first and second layers, $\gamma_3 = -0.413$ eV, $\gamma_4 = -0.177$ eV and $\gamma_5 = -0.119$ eV denote those between the second and third layers, $\gamma_6 = -0.013$ eV, $\gamma_7 = -0.0177$ eV and $\gamma_8 = -0.0119$ eV correspond to those between the first and third layers, and $\gamma_9 = -0.012$ eV accounts for the difference in the chemical environment of the A and B atoms. Only γ_1, γ_3 and γ_6 belong to the vertical interlayer atomic interactions.

AAB stacking possesses a unique electronic structure, in which three pairs of conduction and valence bands have special energy dispersions (figure 6.1(a)). The occupied valence bands are somewhat asymmetric to the unoccupied conduction bands about $E_F = 0$ because of the interlayer atomic interactions. In the vicinity of E_F, the first pair of subbands $S_1^{c,v}$ exhibits strong oscillatory energy dispersions. The conduction and valence states start from extreme points of about ±4 meV at the K point. Along the KM or the KΓ direction, they reach the second point of $\sim \pm 58$ meV, further exhibit oscillatory variations at the third point with $\sim \pm$ 4 meV, and then change monotonously. There exist three constant energy contours within ±58 meV and a narrow energy gap of $E_g \sim 8$ meV, which is absent in other stacking configurations. Such peculiar characteristics can be verified by ARPES [11]. The circular loop closest to the K point has a vanishing radius at the band-edge states so that it does not contribute any special structure to the DOS. This is similar to that in a Dirac-cone structure. Specifically, the other two circular loops can be regarded as the 1D energy bands near the band-edge states; that is, they can display the square root asymmetric peaks in the DOS and spectral structures. These will induce very complicated LL energy spectra, so that the low-energy expansion of the effective

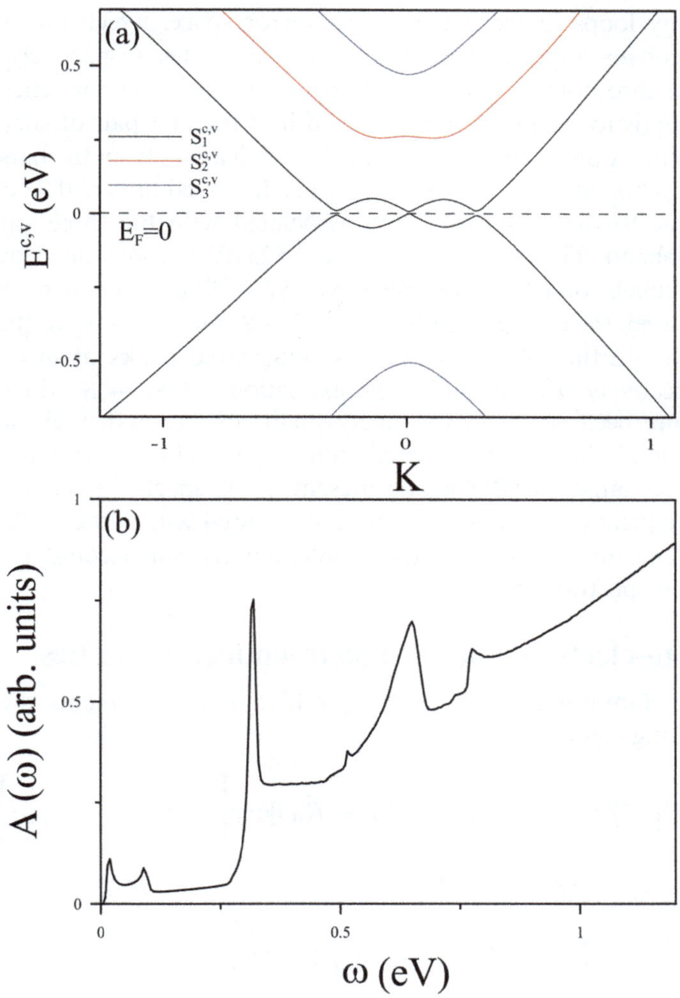

Figure 6.1. (a) The zero-field band structure and (b) the absorption spectrum of the AAB-stacked trilayer graphene.

mass approximation is impossible and will not be able to deal with the magnetic quantization. Concerning the sombrero-shaped bands $S_2^{c,v}$, the energy difference between the two extreme points is quite narrow at only about 5 meV. Away from E_F, the monotonic parabolic bands $S_3^{c,v}$ possess band-edge states at about ±0.49 eV. The specific electronic structures distinguish the AAB stacking from the others. Moreover, the generalized tight-binding model is consistent with the first principle calculations in the energy dispersions (dashed curves) [9].

These unique energy dispersions can enrich the optical properties. There are nine kinds of excitation channel between any valence and conduction band in which plentiful structures are revealed in the absorption spectrum (figure 6.1(b)). At a lower frequency, two asymmetric peak structures appear at $\omega \approx 0.01$ eV and 0.11 eV. This is mainly owing to the fact that each oscillatory subband has three circular constant energy loops in the energy wave vector space, which are related to the concave and convex dispersions near the band-edge states, respectively (figure 6.1).

The optical threshold frequency is determined by the narrow energy gap. The latter corresponds to the oscillation bandwidth of the first pair of energy subbands near the K point. A predominant sharp peak, which arises from the band-edge states of the energy subbands (S_1^v, S_2^c) and (S_2^v, S_1^c) is displayed at $\omega \approx 0.32$ eV. Its strong intensity is due to the very high DOS associated with the three valleys in each oscillatory subband. The weak peak at $\omega \approx 0.525$ eV is contributed by the merged excitation channels of ($S_2^v \rightarrow S_2^c$, $S_1^v \rightarrow S_3^c$, $S_3^v \rightarrow S_1^c$). Moreover, the frequency difference between the optical transitions of $S_2^v \rightarrow S_3^c$ and $S_3^v \rightarrow S_2^c$ in the asymmetric band structures further leads to two distinguishable peaks at $\omega \approx 0.67$ eV and 0.79 eV, respectively. In particular, the excitations of $S_3^v \rightarrow S_3^c$ do not exhibit a special structure because of the extremely small velocity matrix elements. For the middle frequencies, there are many predominant peaks due to the saddle points. The absorption spectrum of AAB stacking presents more spectral structures in the low- and middle-frequency absorption spectrum compared with those of the other types of stacking. The nine interlayer atomic interactions can account for the configuration-created spectral structures.

6.2 Magneto-electronic and magneto-optical properties

The magnetic Hamiltonian with the $12R_B \times 12R_B$ matrix elements is characterized by non-vanishing elements:

$$\langle B_j^1|H|A_i^1\rangle = \gamma_0 \sum \frac{1}{N} \exp[i\vec{k}.(\vec{R}_{A_i'} - \vec{R}_{B_j'})] \exp\left[\frac{2i\pi}{\phi_0}(G_{\vec{R}_{B_j'}} - G_{\vec{R}_{B_j'}})\right] \tag{6.1}$$
$$= \gamma_0 t_{1,i}\sigma_{i,j} + \gamma_0 q\sigma_{i,j+1},$$

$$\langle B_j^2|H|A_i^2\rangle = \gamma_0 t_{1,i}\sigma_{i,j} + \gamma_0 q\sigma_{i,j+1}, \tag{6.2}$$

$$\langle B_j^3|H|A_i^3\rangle = \gamma_0 t_{3,i}\sigma_{i,j-1} + \gamma_0 q\sigma_{i,j}, \tag{6.3}$$

$$\langle A_j^2|H|A_i^2\rangle = \langle A_j^3|H|A_i^3\rangle = \gamma_9\sigma_{i,j}, \tag{6.4}$$

$$\langle A_j^2|H|A_i^1\rangle = \langle B_j^2|H|B_i^1\rangle = \gamma_1\sigma_{i,j}, \tag{6.5}$$

$$\langle B_j^1|H|A_i^2\rangle = \langle A_j^1|H|B_i^2\rangle = \gamma_2 t_{1,i}\sigma_{i,j}, \tag{6.6}$$

$$\langle A_j^3|H|A_i^2\rangle = \gamma_3\sigma_{i,j}, \tag{6.7}$$

$$\langle B_j^3|H|B_i^2\rangle = \gamma_5 t_{2,i}\sigma_{i,j}, \tag{6.8}$$

$$\left\langle B_j^2 \middle| H \middle| A_i^3 \right\rangle = \gamma_4 t_{1,i}\sigma_{i,j} + \gamma_4 q \sigma_{i,j+1}, \tag{6.9}$$

$$\left\langle A_j^2 \middle| H \middle| B_i^3 \right\rangle = \gamma_4 t_{3,i}\sigma_{i,j-1} + \gamma_4 q \sigma_{i,j}, \tag{6.10}$$

$$\left\langle A_j^3 \middle| H \middle| A_i^1 \right\rangle = \gamma_6 \sigma_{i,j}, \tag{6.11}$$

$$\left\langle B_j^3 \middle| H \middle| B_i^1 \right\rangle = \gamma_8 t_{2,i}\sigma_{i,j} + \gamma_8 q \sigma_{i,j+1}, \tag{6.12}$$

$$\left\langle B_j^1 \middle| H \middle| A_i^3 \right\rangle = \gamma_7 t_{1,i}\sigma_{i,j} + \gamma_7 q \sigma_{i,j+1}, \tag{6.13}$$

$$\left\langle A_j^1 \middle| H \middle| B_i^3 \right\rangle = \gamma_7 t_{3,i}\sigma_{i,j-1} + \gamma_7 q \sigma_{i,j}, \tag{6.14}$$

where the four independent phase terms are

$$\begin{aligned}
t_{1,j} &= \exp\{\mathrm{i}[-(k_x b/2) - (\sqrt{3}k_y b/2) + \pi\phi(j-1+1/6)]\} \\
&\quad + \exp\{\mathrm{i}[-(k_x b/2) + (\sqrt{3}k_y b/2) - \pi\phi(j-1+1/6)]\}, \\
t_{2,j} &= \exp\{\mathrm{i}[-(k_x b/2) - (\sqrt{3}k_y b/2) + \pi\phi(j-1+3/6)]\} \\
&\quad + \exp\{\mathrm{i}[-(k_x b/2) + (\sqrt{3}k_y b/2) - \pi\phi(j-1+3/6)]\}, \\
t_{3,j} &= \exp\{\mathrm{i}[-(k_x b/2) - (\sqrt{3}k_y b/2) + \pi\phi(j-1+5/6)]\} \\
&\quad + \exp\{\mathrm{i}[-(k_x b/2) + (\sqrt{3}k_y b/2) - \pi\phi(j-1+5/6)]\}; \\
q &= \exp\{\mathrm{i}k_x b\}.
\end{aligned}$$

The destruction of the inversion symmetry in AAB stacking leads to certain important differences between the LL wavefunctions localized at the 2/6 (5/6) and 1/6 (4/6) centers in terms of LL energy spacings, quantum numbers, spatial distributions and anticrossings. At the 2/6 center, three groups of unoccupied conduction and occupied valence LLs are initiated at 0 (0), −0.23 (0.24) and −0.58 (0.55) eV (figure 6.2).

All the LL wavefunctions are almost well-behaved in the spatial distributions, where the main modes are much stronger than the side modes. Their quantum numbers are determined by the A^1, B^2 and A^2 sublattices, respectively, with the dominating subenvelope functions. For the first group, the ordering of quantum numbers is monolayer-like, in which the middle $n_1^{c,v} = 0$ is located right at E_F. In contrast, $n_2^{c,v}$ begins from $n_2^{c,v} = 1$, then $n_2^{c,v} = 0$, $n_2^{c,v} = 2$, $n_2^{c,v} = 3$, and so on. As for $n_3^{c,v}$, it normally increases from zero for the higher conduction states and the deeper valence states. On the other hand, at the 1/6 center, the quantum numbers of the first LL group can be determined by the B^3 sublattice, while those of the other groups are based on the same sublattices as at the 2/6 center. However, the quantum numbers of the low-lying LLs are not easy to identify from the non-well-behaved spatial distributions, since the complex LL anticrossings cause the subenvelope functions

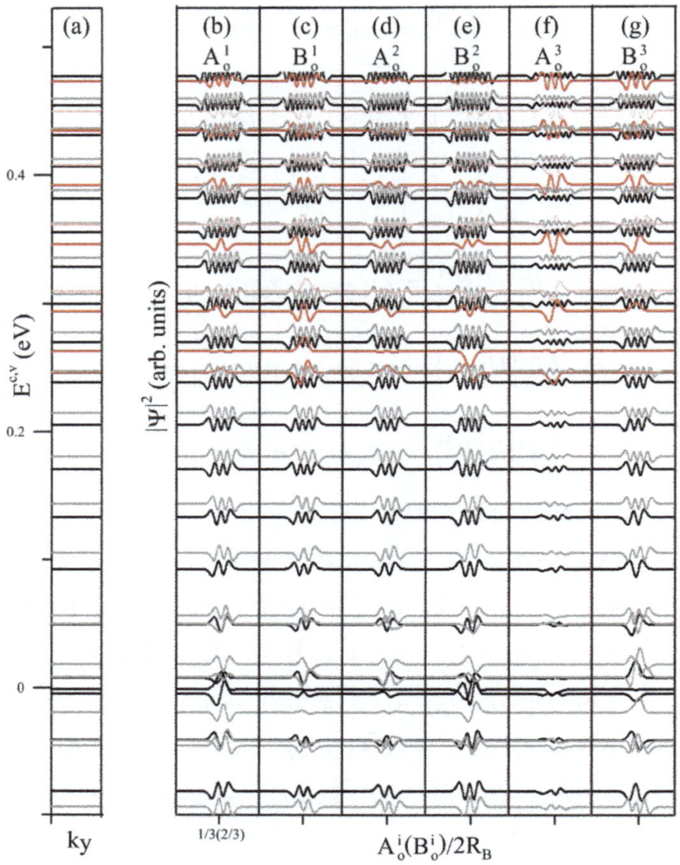

Figure 6.2. The low-lying LL wavefunctions with six distinct sublattices localized at the 2/6 and 4/6 centers of the AAB-stacked trilayer graphene under $B_0 = 40$ T.

to oscillate abnormally. Remarkably, the quantum numbers are out of normal ordering, different from those of monolayer graphene. The LL splitting in AAB stacking is expected to diversify the magneto-absorption spectra.

The AAB stacking configuration induces special relationships between the different subenvelope functions in amplitudes and phases. The amplitudes of the six subenvelope functions, which depend on the interlayer atomic interactions, are non-equivalent. Explicitly, at the 2/6 center, the amplitude of the A^3 sublattice is much weaker than that of the others in the first LL group (figure 6.2). Similar results are also revealed for the A^2 and B^2 sublattices in the second group as well as the B^3 sublattice of the low-lying LLs in the third group. Moreover, the A^1 and A^2 (B^1 and B^2) sublattices have similar subenvelope functions with the same number of zero points, mainly owing to the same (xy) projections. Concerning the phases in the first and second LL groups, those of the A^1 and A^2 sublattices are identical to all the conduction states but have distinct signs for the valence states in the first and second LL groups. The opposite is true for the third LL group. These phase relationships can also be observed for the B^1 and B^2 sublattices in the first and third groups, but are opposite in the second group. The unique features of the subenvelope functions can greatly enrich the absorption spectra.

AAB stacking reveals the feature-rich B_0-dependent LL energy spectrum. Three groups of LL energies exhibit monotonic and entangled B_0-dependence. The conduction and valence LL energies monotonously approach the initial energies upon decreasing the field strength from $B_0 = 60$ T (figure 6.3). However, for the first and second groups, unusual n^{cv} ordering and B_0-dependence occur below the critical magnetic fields. There exists an oscillatory B_0-dependent energy spectrum, directly reflecting the strong competition in the magnetic quantization of multicontour energy loops (figure 6.1). Furthermore, a simple relationship between the LL energy spacing and B_0 or $n_1^{c,v}$ is absent. Specifically, diverse anticrossing and crossing phenomena are revealed in distinct ($E^{c,v}$, B_0) regions including the intragroup LL anticrossings/crossings from the first and the second groups (at low B_0), and the intergroup LL anticrossings/crossings between any two groups (at $B_0 > 80$ T; not shown). The main modes of the two anticrossing LLs have a difference of $\Delta n = \pm 3i$ (i is an integer). These special features are expected to induce unique magneto-absorption frequency spectra.

Three intragroup and six intergroup optical excitations are clearly revealed in the spectra, as indicated in figure 6.4. The excitations are in sharp contrast with each other in the peak structures, spectral intensities and energy spacings in the twin peaks. The peak structures are determined by the quantum modes, LL energy spectra and fermion distribution. In general, the intragroup excitations present very weak twin peaks. In contrast, the intergroup absorption peaks are single peaks with monolayer-like strong intensities. Exceptionally, the transitions between the first and

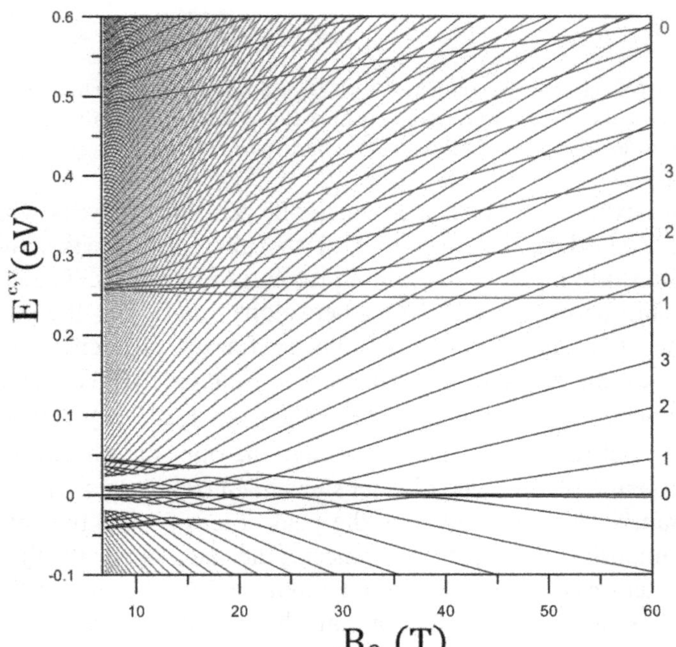

Figure 6.3. The Landau-level spectrum of AAB-stacked trilayer graphene.

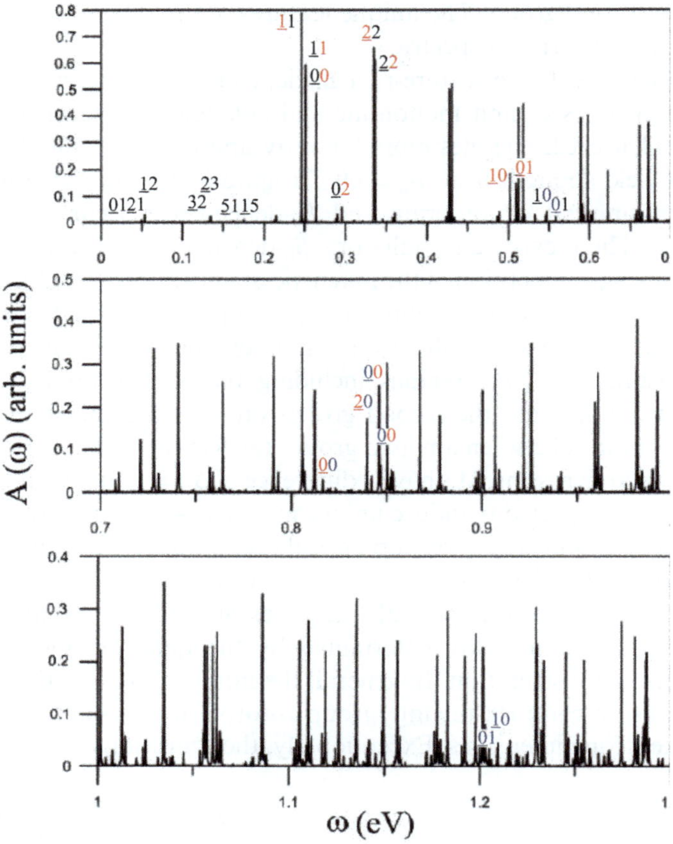

Figure 6.4. The magneto-absorption spectrum of AAB-stacked trilayer graphene localized at the 1/6 center with three intragroup and six intergroup categories of LL excitations under $B_0 = 40$ T. The black, red and blue numbers, respectively, indicate the excitations from the first, second and third LL groups.

third LL groups exhibit twin peaks, in which one peak is very weak while another is rather strong. Regarding the intensities of the absorption peaks, the intergroup optical excitations are much stronger than the intragroup ones. The amplitude and phase relationship between the A^l (B^l) subenvelope function of the initial state and the B^l (A^l) one of the final state is responsible for the differences in the spectral intensities. For example, the intragroup peak $\omega_1^{1^v 2^c}$ at $\omega \approx 0.06$ eV is about 1/15 of the intergroup one $\omega_{21}^{1^v 1^c}$ at $\omega \approx 0.25$ eV. For the former, the $|B^1\rangle$ of the $n_1^v = 1$ initial state and the $\langle A^1|$ of the $n_1^c = 2$ final state have the opposite phases, while $|B^2\rangle$ and $\langle A^2|$ possess the same phases. The distinct signs in $\langle A^1|B^1\rangle$ and $\langle A^2|B^2\rangle$ greatly reduce the velocity matrix elements and thus the peak intensity. On the other hand, the strong peak intensity in the latter reflects the fact that all the dominating subenvelope functions of the $n_2^v = 1$ initial and $n_1^c = 1$ final states on the sublattices, (B^1, A^1), (B^2, A^2) and (A^3, B^3), have similar phase relationships.

At the 2/6 center, the intragroup inter-LL excitations possess a lot of low-intensity absorption peaks which mostly belong to the twin-peak structure (figure 6.5).

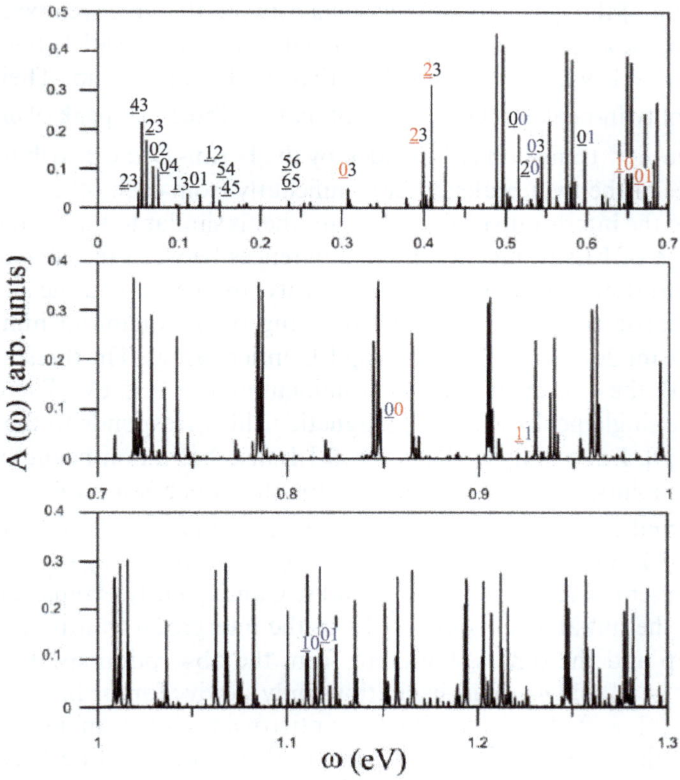

Figure 6.5. The magneto-absorption spectrum of AAB-stacked trilayer graphene localized at the 2/6 center with three intragroup and six intergroup categories of LL excitations under $B_0 = 40$ T. The black, red and blue numbers, respectively, indicate the excitations from the first, second and third LL groups.

Exclusively, the threshold peak of $\omega_1^{0^v1^c}$ ($\omega \approx 0.01$ eV) is a single peak with ignorable intensity due to the extremely weak subenvelope functions of $n_1^v = 0$ LL in the B^1, A^2 and A^3 sublattices. Distinctively, the threshold peaks of ($\omega_2^{0^v1^c}$, $\omega_2^{1^v0^c}$) and ($\omega_3^{0^v1^c}$, $\omega_3^{1^v0^c}$) are twin peaks in which the former are relatively strong while the latter are comparable with the other intragroup excitation peaks. This is because all the subenvelope functions of the $n_2^v = 0$ (1) initial and $n_2^c = 1$ (0) final states on the A^l and B^l sublattices have the same (opposite) phases. In general, three categories of intragroup inter-LL excitations exhibit similar characteristics, including peak intensity, twin-peak structure and selection rule ($\Delta n = \pm 1$).

For the intergroup inter-LL excitations, besides the asymmetry of the LL energy spectra, the quantum modes are responsible for the peak structures. Specifically, the absorption peaks associated with the second LL group are single peaks. Concerning the $n_1^v \to n_2^c$ and $n_2^v \to n_1^c$ transitions, the initial and final states of the available excitations satisfy $n_1^{c,v} = n_2^{v,c}$. Remarkably, the threshold peaks of $\omega_{1,2}^{0^v0^c}$ and $\omega_{2,1}^{0^v0^c}$ merge together at $\omega = 0.265$ eV and become a higher-intensity single peak. As for the $n_2^v \to n_3^c$ and $n_3^v \to n_2^c$ excitations, the single peaks mainly come from the inter-LL transitions with $|n_2^{c,v} - n_3^{v,c}| = 2$ and 0, in which the former are much higher than the latter as a result of the special phase relations between the subenvelope functions. On the other hand, the $n_3^v \rightarrow n_1^c$ and $n_1^v \rightarrow n_3^c$ inter-LL transitions satisfy the selection rule of $|n_1^{c,v} - n_3^{v,c}| = 1$, which is identical to that of the intragroup. Their absorption spectra present twin-peak structures, except that the threshold peak of $\omega_{3,1}^{0^v1^c}$ is a single form, since the $\omega_{3,1}^{1^v0^c}$ transition is forbidden by the Fermi–Dirac distribution function. The intensities of the twin peaks differ significantly from each other. One of them is comparable to the intergroup peaks, while another is similar to that of the intragroup.

The splitting of LLs results in critical differences between the 1/6- and 2/6 center magneto-absorption spectra in terms of intensity, frequency and the structure of the peaks. These arise from the abnormal ordering of the quantum number and the significant side modes during the low-lying LL anticrossings. The threshold peak $\omega_1^{0^v1^c}$ associated with the 1/6 center is quite prominent at $\omega \approx 0.15$ eV. There are a lot of low-frequency single peaks when the magnetic field corresponds to the range of LL anticrossings [9]. For example, the $n_1^{c,v} = 0$ LL also has a non-negligible side mode of three zero points, which is responsible for the anticrossing of the $n_1^{c,v} = 0$ and $n_1^{c,v} = 3$ LLs and the appearance of the single peaks $\omega_1^{0^v2^c}$ and $\omega_1^{0^v4^c}$. Furthermore, many pairs of LLs in the anticrossings such as $(n_1^{c,v} = 1, n_1^{c,v} = 4)$, $(n_1^{c,v} = 2, n_1^{c,v} = 5)$, and so on, present several extra single peaks. Consequently, similar characteristics also occur in the initial absorption peaks of the intergroup transitions between the first LL group and the others. Different from the absorption spectrum at the 2/6 center, the $n_2^v \rightarrow n_3^c$ and $n_3^v \rightarrow n_2^c$ excitations only survive for the inter-LL transitions with $|n_2^{c,v} - n_3^{v,c}| = 0$. In short, the LL anticrossings play an important role in distinguishing the magneto-absorption spectra due to the splitting LLs.

The optical absorption peak frequencies of AAB stacking exhibit the plentiful B_0-dependent spectrum. Both the intra- and intergroup excitation frequencies have non-monotonic B_0-dependence except for the $3^v \rightarrow 3^c$ intragroup excitation category. As B_0 gradually decreases from 60 T, the absorption frequencies normally decline. However, they start to grow after certain critical magnetic fields, such as $\underline{2}1$ at $B_0 \approx 26$ T and $\underline{1}1$ at $B_0 \approx 38$ T (figure 6.6(a) and (b)). This is associated with the abnormal field-dependent LL energy spectrum magnetically quantized from the oscillatory and the sombrero-shaped energy bands. It should be noted that LL anticrossing phenomena are also present in discontinuous absorption frequencies at certain B_0, e.g. $(\underline{2}1, \underline{1}2)$ at $B_0 \approx 18$ T and $(\underline{1}1, \underline{1}1)$ at $B_0 \approx 25$ T (figure 6.6(a) and (b)). An unusual absorption frequency spectrum is easily observed at low field strength, as indicated by more absorption peaks. Specifically, the excitation channels of the threshold peaks change dramatically from $\underline{0}1$ to $\underline{0}2$, $\underline{4}0$, $\underline{5}0$ and others, being attributed to the frequent intragroup LL anticrossings (figure 6.3). With regard to the intergroup transitions, the initial peaks, $\underline{0}0$ and $\underline{0}0$, are forbidden for B_0, respectively, smaller and larger than 17 T, by the requirement of the Fermi–Dirac distribution. Furthermore, they have weakly dispersive relations between ω_a and B_0, since the LL energies of $n_1^{v,c} = 0$ and $n_2^{c,v} = 0$ are almost initiated at specific energies. Significantly, quantum number ordering plays an important role in the B_0 dependence. The intergroup threshold peak, in the range of $28\text{ T} < B_0 < 51$ T, is not the initial peak, but is replaced by the second excitation channel $\underline{1}1$. This characteristic comes from the unusual $n_2^{c,v}$ ordering

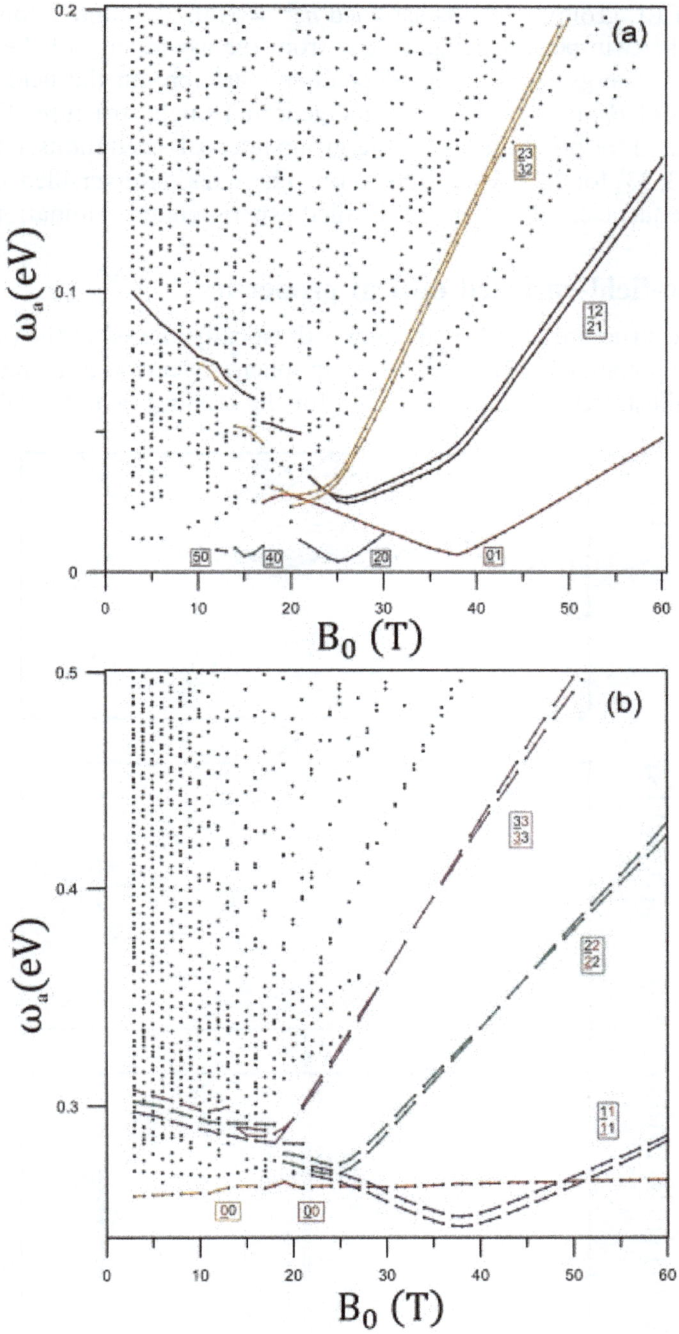

Figure 6.6. The magneto-absorption spectrum of AAB-stacked trilayer graphene localized at the 1/6 center with three intragroup and six intergroup categories of LL excitations under $B_0 = 40$ T. The black, red and blue numbers, respectively, indicate the excitations from the first, second and third LL groups.

of the second LL group, which is assigned $n_2^{v,c} = 1, 0, 2, 3$, and so on (figure 6.3). Concerning the twin-peak spacing arising from the asymmetry of the valence and conduction LL energy spectra, this depends remarkably on the field strength. At certain ranges of B_0, twin peaks might merge into a single structure. This feature is clearly illustrated for both intra- and intergroup inter-LL excitations, e.g. ($\underline{1}2$, $\underline{2}1$) for $B_0 < 24$ T, ($\underline{3}3$, $\underline{3}3$) for $B_0 < 14$ T, and so on. The stacking-diversified B_0-dependent absorption frequencies are worthy of detailed experimental examination [12–35].

6.3 Electric-field-enriched optical properties

The electronic structure of AAB stacking is drastically altered in the presence of **E**. The Coulomb potential difference between graphene layers has a strong effect on the band gap as illustrated in figure 6.7(a). As for the oscillatory and sombrero-shaped

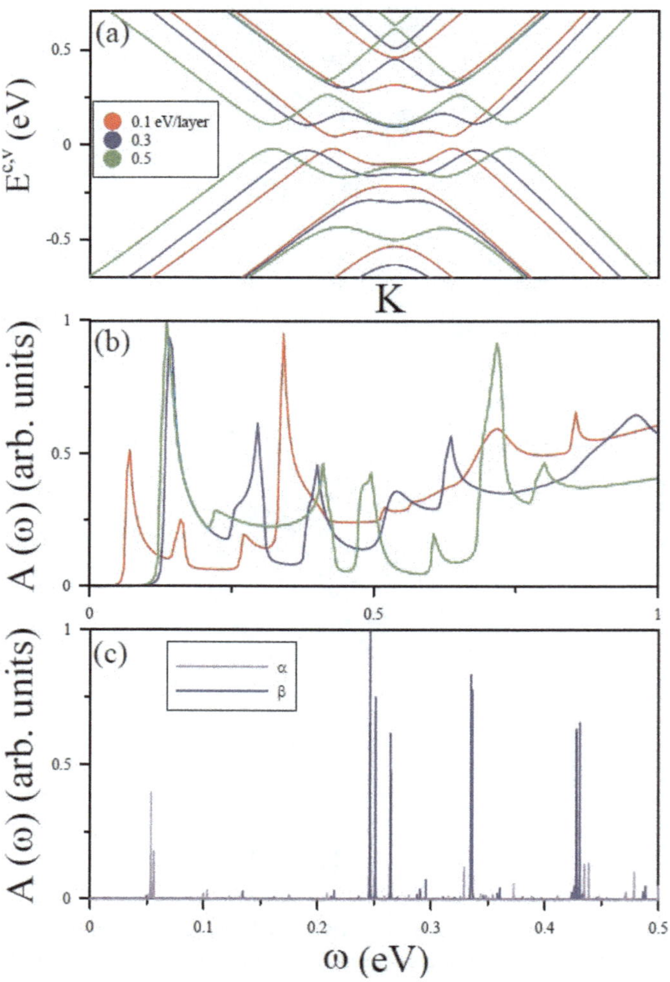

Figure 6.7. The E_0-enriched (a) band structure, (b) absorption and (c) magneto-absorption spectra of AAB-stacked trilayer graphene.

energy bands, their oscillation widths are greatly widened, and the band-edge states move away from the K point. This induces larger constant energy loops, indicating higher DOS, and the energy spacings of the concave and convex energy dispersions become large. The parabolic subbands might change into linear ones when the electric field is sufficiently large. Remarkably, the asymmetry between the valence and conduction subbands is greatly amplified by the close cooperation of the complex interlayer atomic interactions and the electric field. Furthermore, their constant energy loops (or band-edge states) correspond to the different wave vectors.

The frequency, intensity and number of the spectral structures are very sensitive to changes in the electric field strength, as shown in figure 6.7(b). With the increase of E_0, the optical gap grows monotonously, and then reaches a saturated value at higher fields ($E_0 \geqslant 0.3$ eV/layer; blue and green curves). The intensity of the threshold peak is greatly enhanced as a result of the E_0-enlarged constant energy loops. A similar result is also revealed in the second peak. There exists an extra weaker peak between these two prominent ones, reflecting the distinct band-edge states in the valence and conduction bands. Moreover, the original merged absorption peaks might have a clear separation, such as the splitting from one peak into two, related to $S_1^v \rightarrow S_2^c$ and $S_2^v \rightarrow S_1^c$. This is caused by the E_0-enhanced asymmetric energy spectrum.

The double degeneracy of LLs in AAB stacking remains the same even under an electric field, i.e. the doubly degenerate LLs are characterized by the 1/6 and 2/6 localization centers in a composite field. However, the electric field can amplify the asymmetry of the LL energy spectra, alter the spatial distributions in distinct sublattices and generate more frequent anticrossings, leading to drastic changes in the frequency, intensity and number of magneto-absorption peaks. There are more observable symmetric peaks as a result of the unmerged inter-LL excitation channels and the side-mode-induced ones, as shown in figure 6.7(c) at $E_0 = 0.1$ eV/layer and $B_0 = 40$ T for the 2/6 case.

The electronic and optical properties of AAB stacking are easily modulated by external fields. Experimental observations of this have been absent up to now. The unique E_0 dependence of the band-edge state energies can be directly verified through the STS [36] and ARPES measurements [37]. STS is useful in examining the B_0-dependent splitting LL energy spectra with strong asymmetry about E_F, the frequent intra- and intergroup anticrossings and the complex E_0 dependence. Moreover, many spectral structures at the low and middle frequency, magneto-absorption peaks with distinct structures, the diverse selection rules and the E_0-enriched optical spectra need further examination using various optical spectroscopies [37].

6.4 Comparisons between the different types of trilayer stacking

The stacking configurations can diversify energy dispersions and the optical properties of trilayer graphene. As for the low-lying band structure, there are critical differences between the four stacking configurations. The AAA- and ABA-stacked trilayer graphene possess special band structures with three pairs of linear bands and a pair of linear bands, and two pairs of parabolic bands, respectively. Their electronic structures can be, respectively, regarded as the superposition of monolayer graphene, and monolayer and bilayer graphene. In contrast, the ABC and AAB stacking exhibit unique energy bands, which are absent in monolayer and bilayer systems. The former has pairs of partially flat, sombrero-shaped and linear bands, and the latter possesses pairs of oscillatory, sombrero-shaped and parabolic bands. Remarkably, only the AAB stacking is a narrow direct-gap semiconductor, while the others are gapless 2D semimetals.

The diverse energy dispersions are directly reflected in the rich optical spectrum. For the AAA and ABA stacking, only the intra-Dirac-cone optical excitations can survive as a result of the linearly symmetric or anti-symmetric superposition of the wavefunctions in different layers. There is solely a shoulder structure in the low-frequency absorption spectra. As for the former, it is displayed right after the forbidden excitation region under specific fermion distributions due to the almost symmetric upper and lower Dirac cones. However, that of the latter comes from the distinct parabolic bands. In the middle-frequency spectrum, these two systems, respectively, present three and five π-electronic absorption peaks coming from the saddle-points of three pairs of subbands. Regarding ABC and AAB stacking, excitation channels between any valence and conduction band are available. The threshold frequency of the former is zero, while that of the latter is determined by the narrow energy gap. The low-frequency spectra exhibit peak and shoulder structures, which are different from the two above-mentioned stacking systems. AAB stacking presents the most plentiful spectral structures due to the complex interlayer atomic interactions. Under an electric field, the spectral characteristics of the AAA stacking remain similar, but the opposite is true for the other ones. ABA stacking has 3×3 excitation channels, as observed in ABC and AAB stacking. More special absorption structures are created in these three systems; furthermore, an optical gap is opened for ABC stacking.

The magneto-absorption spectra are greatly enriched by the stacking configurations. The AAA, ABA, ABC and AAB stacking sharply contrast with one another in the excitation category, structure, intensity and number of absorption peaks. This is mainly determined by the special relationships between the sub-envelope functions on the six sublattices, the Fermi–Dirac distribution, the symmetric or asymmetric LL energy spectra and the LL anticrossings. The AAA and ABA stacking only present three and five categories of absorption peaks, respectively. The linearly symmetric and antisymmetric superposition of the LL wave-functions in different layers can account for the absence of intergroup excitations related to the Dirac cone LLs. On the other hand, the ABC- and AAB-stacked trilayer graphene possess nine categories of absorption peaks. Regarding spectral form and intensity, the intra-Dirac-cone LL excitations in the AAA and ABA stacking are mainly present in the single peaks with a predominantly uniform intensity, owing to the almost symmetric LL spectra and same velocity matrix elements. Specifically, the ABA stacking also has a bilayer-like magneto-absorption spectrum in the twin-peak structure. The ABC stacking exhibits twin-peak structures except for the initial absorption peaks. In general, the intragroup excitations are stronger than the intergroup ones, which is in sharp contrast with the AAB stacking. The AAB stacking displays more complex magneto-absorption spectra with many single and twin peaks, reflecting the complex relationship between the interlayer atomic interactions and the magnetic quantization. The intensity difference between the intra- and intergroup excitation peaks is more obvious. In particular, the LL anticrossings and the splitting LLs at different localization centers significantly increase the number of absorption peaks. The critical differences between the four stacking systems can provide useful information in identifying the specific stacking configurations by the experimental measurements [37].

The magneto-absorption spectra can be diversified by an electric field, depending on the stacking configurations. The main features of the optical properties in AAA stacking hardly change under a composite external field, while those of the other three are greatly enriched. The obvious changes of the magneto-absorption peaks lie in frequency and intensity. In ABA and ABC stacking, extra peaks and double peaks are generated, respectively, by the LL anticrossings and splitting LLs. In particular, the former displays more inter-LL excitation categories which are absent in the other stacking systems.

References

[1] Rong Z Y and Kuiper P 1993 Electronic effects in scanning tunneling microscopy: Moiré pattern on a graphite surface *Phys. Rev.* B **48** 17427

[2] Xu P, Xu P, Yang Y, Qi D, Barber S D, Schoelz J K and Ackerman M L *et al* 2012 Electronic transition from graphite to graphene via controlled movement of the top layer with scanning tunneling microscopy *Phys. Rev.* B **86** 085428

[3] Rong Z Y 1994 Extended modifications of electronic structures caused by defects: scanning tunneling microscopy of graphite *Phys. Rev.* B **50** 1839–43

[4] Pong W T, Bendall J and Durkan C 2007 Observation and investigation of graphite superlattice boundaries by scanning tunneling microscopy *Surf. Sci.* **601** 498

[5] Campanera J M, Savini G, Suarez-Martinez I and Heggie M I 2007 Density functional calculations on the intricacies of Moiré patterns on graphite *Phys. Rev.* B **75** 235449

[6] Kazemi A S, Crampin S and A Ilie 2013 Stacking-dependent superstructures at stepped armchair interfaces of bilayer/trilayer graphene *Appl. Phys. Lett.* **102** 163111

[7] Biedermann L B, Bolen M L, Capano M A, Zemlyanov D and Reifenberger R G 2009 Insights into few-layer epitaxial graphene growth on 4H-SiC(000 $\bar{1}$) substrates from STM studies *Phys. Rev.* B **79** 125411

[8] Y Que, Xiao W, Chen H, Wang D, S Du and Gao H J 2015 Stacking-dependent electronic property of trilayer graphene epitaxially grown on Ru(0001) *Appl. Phys. Lett.* **107** 263101

[9] Do T N, Lin C Y, Lin Y P, Shih P H and Lin M F 2015 Configuration-enriched magneto-electronic spectra of AAB-stacked trilayer graphene *Carbon* **94** 619–32

[10] Do T N, Shih P H, Chang C P, Lin C Y and Lin M F 2015 Rich magneto-absorption spectra in AAB-stacked trilayer graphene *Phys. Chem. Chem. Phys* **18** 17597–605

[11] Sprinkle M, Siegel D, Hu Y, Hicks J, Tejeda A and Taleb-Ibrahimi A *et al* 2009 First direct observation of a nearly ideal graphene band structure *Phys. Rev. Lett.* **103** 226803

[12] Gusynin V P, Sharapov S G and Carbotte J P 2007 Anomalous absorption line in the magneto-optical response of graphene *Phys. Rev. Lett.* **98** 157402

[13] Li Z Q, Henriksen E A, Jiang Z, Z Hao, Martin M C and Kim P *et al* 2008 Dirac charge dynamics in graphene by infrared spectroscopy *Nat. Phys.* **4** 532–5
[14] Zhang L M, Li Z Q, Basov D N, Fogler M M, Hao Z and Martin M C 2008 Determination of the electronic structure of bilayer graphene from infrared spectroscopy *Phys. Rev.* B. **78** 235408
[15] Kuzmenko A B, van Heumen E, van der Marel D, Lerch P, Blake P and Novoselov K S *et al* 2009 Infrared spectroscopy of electronic bands in bilayer graphene *Phys. Rev.* B. **79** 115441
[16] Li Z Q, Henriksen E A, Jiang Z, Hao Z, Martin M C and Kim P *et al* 2009 Band structure asymmetry of bilayer graphene revealed by infrared spectroscopy *Phys. Rev. Lett.* **102** 037403
[17] Mak K F, Shan J and Heinz T F 2010 Electronic structure of few-layer graphene: experimental demonstration of strong dependence on stacking sequence *Phys. Rev. Lett.* **104** 176404
[18] Jiang Z, Henriksen E A, Tung L C, Wang Y J, Schwartz M E and M Y Han *et al* 2007 Infrared spectroscopy of Landau levels of graphene *Phys. Rev. Lett.* **98** 197403
[19] Daboussi A, Mandhour L, Fuchs J N and Jaziri S 2014 Tunable zero-energy transmission resonances in shifted graphene bilayer *Phys. Rev.* B **89** 085426
[20] Plochocka P, Faugeras C, Orlita M, Sadowski M L, Martinez G and Potemski M *et al* 2008 High-energy limit of massless Dirac fermions in multilayer graphene using magneto-optical transmission spectroscopy *Phys. Rev. Lett.* **100** 087401
[21] Kim D S, Kwon H, Nikitin A Y, Ahn S, Martin-Moreno L and Garcia-Vidal F J *et al* 2015 Stacking structures of few-layer graphene revealed by phase-sensitive infrared nanoscopy *ACS Nano* **9** 6765–73
[22] Faugeras C, Amado M, Kossacki P, Orlita M, Kuhne M and Nicolet A A L *et al* 2011 Magneto-Raman scattering of graphene on graphite: electronic and phonon excitations *Phys. Rev. Lett.* **107** 036807
[23] Zhang X, Han W P, Qiao X F, Tan Q H, Wang Y F and Zhang J *et al* 2016 Raman characterization of AB- and ABC-stacked few-layer graphene by interlayer shear modes *Carbon* **99** 118–22
[24] Raghavan S, Denig T J, Nelson T C, Chaudhari S and Stinespring C D 2016 Characterization of graphene-on-insulator films formed using plasma based surface chemistry *Carbon* **99** 212–21
[25] Ferrari A C, Meyer J C, Scardaci V, Casiraghi C, Lazzeri M and Mauri F *et al* 2006 Raman spectrum of graphene and graphene layers *Phys. Rev. Lett.* **97** 187401
[26] Nyakiti L O, Myers-Ward R L, Wheeler V D, Imhoff E A, Bezares F J and Chun H *et al* 2012 Bilayer graphene grown on 4H-SiC (0001) step-free mesas *Nano Lett.* **12** 1749–56
[27] Lui C H, Li Z, Chen Z, Klimov P V, Brus L E and Heinz T F 2011 Imaging stacking order in few-layer graphene *Nano Lett.* **11** 164–9
[28] Cancado L G, Takai K, Enoki T, Endo M, Kim Y A and Mizusaki H *et al* 2008 Measuring the degree of stacking order in graphite by Raman spectroscopy *Carbon* **46** 272–5
[29] Zhao H M, Lin Y C, Yeh C H, Tian H, Chen Y C and Xie D *et al* 2014 Growth and Raman spectra of single-crystal trilayer graphene with different stacking orientations *ACS Nano* **8** 10766–73
[30] Berciaud S, Potemski M and Faugeras C 2014 Probing electronic excitations in mono- to pentalayer graphene by micro magneto-Raman spectroscopy *Nano Lett.* **14** 4548–53

[31] Faugeras C, Binder J, Nicolet A A L, Leszczynski P, Kossacki P and Wysmolek A *et al* 2014 A micro-magneto-Raman scattering study of graphene on a bulk graphite substrate *EPL* **108** 27011
[32] Fang W, Hsu A L, Caudillo R, Song Y, Birdwell A G and Zakar E *et al* 2013 Rapid identification of stacking orientation in isotopically labeled chemical-vapor grown bilayer graphene by Raman spectroscopy *Nano Lett.* **13** 1541–8
[33] Poncharal P, Ayari A, Michel T and Sauvajol J L 2008 Raman spectra of misoriented bilayer graphene *Phys. Rev.* B **78** 113407
[34] Malard L M, Nilsson J, Elias D C, Brant J C, Plentz F and Alves E S *et al* 2007 Probing the electronic structure of bilayer graphene by Raman scattering *Phys. Rev.* B **76** 201401
[35] Xie L M, Wang H L, Jin C H, Wang X R, Jiao L Y and Suenaga K *et al* 2011 Graphene nanoribbons from unzipped carbon nanotubes: atomic structures, Raman spectroscopy, and electrical properties *J. Am. Chem. Soc.* **133** 10394–7
[36] Cherkez V, Trambly de Laissardiere G, Mallet P and Veuillen J Y 2015 Van Hove singularities in doped twisted graphene bilayers studied by scanning tunneling spectroscopy *Phys. Rev.* B **91** 155428
[37] Ohta T, Bostwick A and McChesney J L 2007 Interlayer interaction and electronic screening in multilayer graphene investigated with angle-resolved photoemission spectroscopy *Phys. Rev. Lett.* **98** 206802

IOP Publishing

Optical Properties of Graphene in Magnetic and Electric Fields

Chiun-Yan Lin, Thi-Nga Do, Yao-Kung Huang and Ming-Fa Lin

Chapter 7

Sliding bilayer graphene

In addition to high-symmetry stacking, low-symmetry stacking also deserves systematic research [1–14]. Sliding bilayer graphene (BLG), with various misalignment configurations, is chosen for a detailed discussion about its specific essential properties. During the shift between two graphene layers, the first principle calculations show that the variation of the total ground state energy is small in configuration-dependent BLG [15]. Moreover, some recent experimental studies have successfully synthesized asymmetric stacking using specific techniques [16]. Stacking boundaries in which one of the graphene layers is shifted with respect to another have been successfully grown using the CVD method [17]. The sliding of graphene flakes on a graphene substrate can also be initiated by the STM tip to overcome the weak van der Waals interactions [16]. In addition, micrometer-sized graphite flakes can slide spontaneously after being stirred by an STM tip [18, 19].

The generalized tight-binding model is suitable for studying magnetic quantization in high- and low-symmetry BGL. The variation in stacking configuration can induce a dramatic transformation between the Dirac cone structure and the parabolic band structure [15, 20–23], and thus create unusual absorption spectra in terms of the optical gap and low- and middle-frequency special structures. The low-frequency optical properties are significantly modified by electric fields. Moreover, the LLs of sliding BLG exhibit drastic changes in the spatial symmetry, initial formation energy, intergroup anticrossings, state degeneracy and semiconductor–metal transition. There are three kinds of LLs: well-behaved, perturbed and undefined, which are characterized by a specific mode, a main mode plus side modes, and a disordered mode, respectively. Such LLs are clearly revealed in the diverse magneto-optical selection rules. In particular, the undefined LLs frequently present intergroup anticrossings in the B_0-dependent energy spectra, and exhibit a large number of absorption peaks in the absence of selection rules. Diverse selection rules, the special B_0-dependent energy spectra, and the unusual zero-field absorption spectra are closely related to the stacking configurations; therefore, their experimental measurement can be used to resolve the subangstrom misalignment stacking of BLG.

doi:10.1088/978-0-7503-1566-1ch7

© IOP Publishing Ltd 2017

7.1 Electronic and optical properties without external fields

The distinct stacking configurations of BLG can be tuned by the relative shift between two layers. When a shift happens along the armchair direction in the range of $0 \leqslant \delta \leqslant 12b/8$, the stacking configurations are altered in the high-symmetry graphene: AA ($\delta = 0$) $\longrightarrow$ AB ($\delta = b$) $\longrightarrow$ AA′ ($\delta = 1.5b$) (figure 7.1). The C atoms in the AA′ stacking have distinct xy projections, while they experience the same chemical environment, as observed in AA stacking. Four carbon atoms are included in a primitive unit cell. For all the stacking configurations [15, 24, 25], the low-energy electronic properties are dominated by the $2p_z$ orbitals, as indicated by the agreement between the tight-binding model and the first principle calculations. There exist complicated interlayer atomic interactions during the variation of the shift. A position-dependent empirical formula is chosen to characterize the intra- and interlayer atomic interactions:

$$-\gamma_{ij} = \gamma_0 e^{-\frac{d-b}{\varrho}}\left[1 - \left(\frac{\mathbf{d} \cdot \hat{z}}{d}\right)^2\right] + \gamma_1 e^{-\frac{d-d_0}{\varrho}}\left(\frac{\mathbf{d} \cdot \hat{z}}{d}\right)^2 \tag{7.1}$$

where γ_{ij} is the atomic interaction between the lattice sites i and j, depending on the distance and the angle of the two $2p_z$ orbitals. $\gamma_0 = -2.7$ eV is the nearest-neighbor hopping integral in the same layer, $\gamma_1 = 0.48$ eV is the interlayer vertical hopping integral, $\mathbf{d}$ is the position vector connecting two lattice points, $d_0 = 3.35$ Å is the interlayer distance and $\varrho = 0.184\ b$ is the decay length. Equation (7.1) is acceptable for describing the interlayer atomic interactions in carbon-related systems, e.g. multiwalled carbon nanotubes [24, 25].

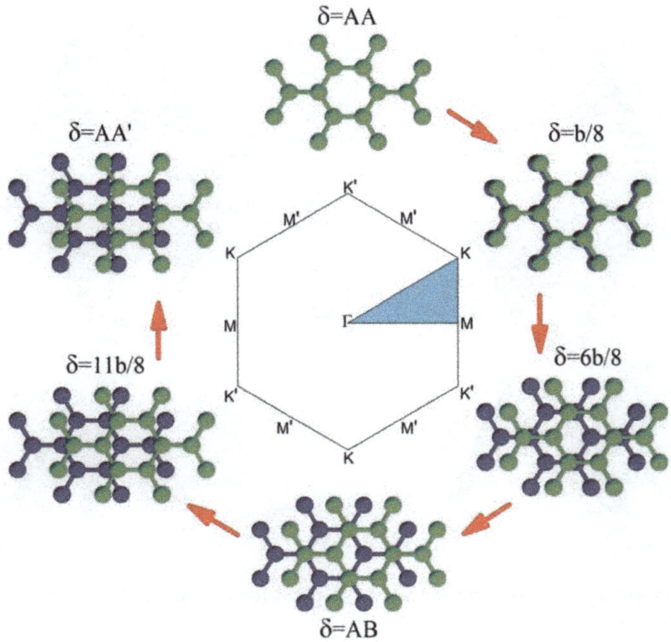

Figure 7.1. The geometric structures of shift-dependent bilayer graphene along the armchair direction.

Sliding BLG exhibits dramatic changes in the band structures. The different stacking systems possess overlaps in the valance and conduction bands (figure 7.2(a)–(f)); that is, they are semimetallic systems with a low free carrier density. The space-inversion symmetry stays the same in the presence of a shift so that each valley in the first Brillouin zone is identical [26]. The K valley is sufficient for a model study. With a small shift ($\delta = b/8$ in figure 7.2(b)), two pairs of isotropic Dirac cones in the AA stacking (figure 7.2(a)) change into distorted energy dispersions, and an eye-shaped

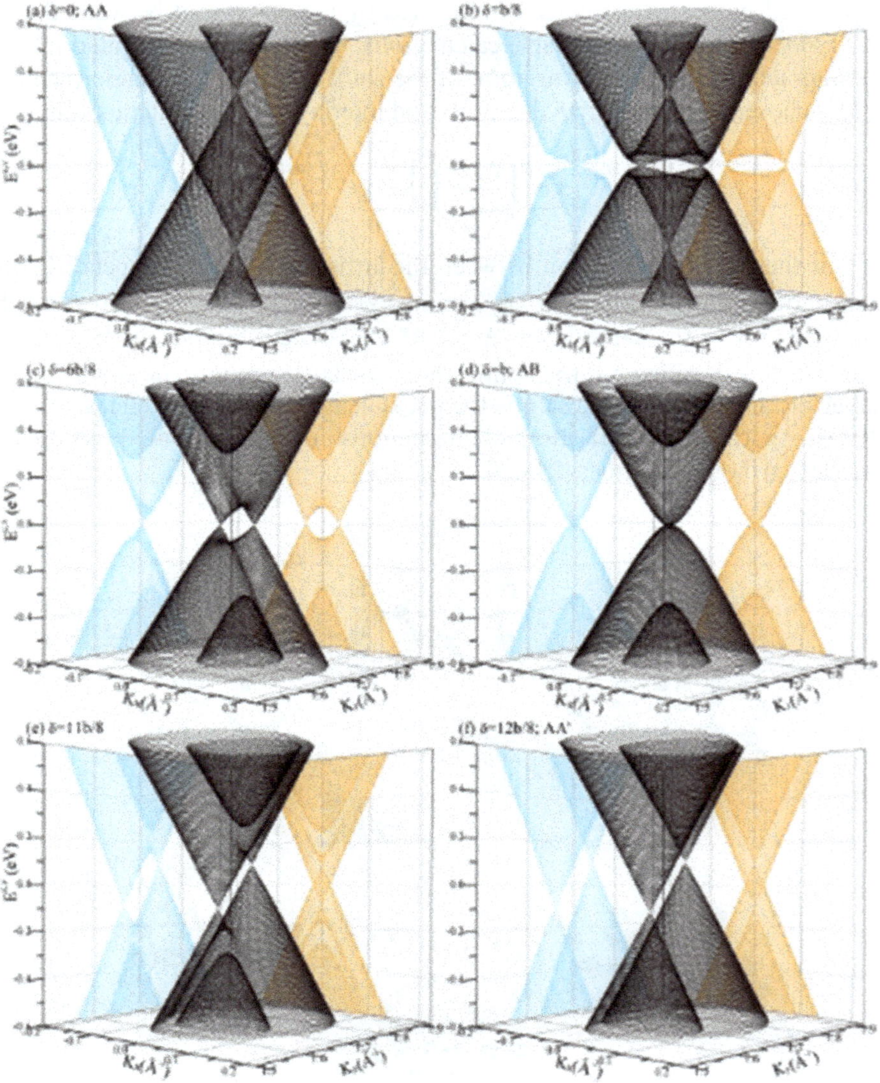

Figure 7.2. The low-energy band structures near the K point for various stacking configurations: (a) $\delta = 0$, (b) $b/8$, (c) $6b/8$, (d) b, (e) $11b/8$ and (f) $12b/8$.

stateless region is created near the Fermi level. The band structure is symmetric about the k_x-axis, since the $y \rightarrow -y$ and space-inversion symmetries are unbroken. A pair of band contact points remain as Fermi momentum states. The Dirac cone structures gradually become smoother with the further increase of δ. After serious distortion, the Dirac points are fully separated near $\delta \cong 6b/8$ (figure 7.2(c)).

A further shift in the range of $b \leqslant \delta \leqslant 1.5b$ also leads to the special transformation of energy dispersions. Two pairs of parabolic bands are formed in the AB stacking of $\delta = b$ (Figure 7.2(d)), as verified by ARPES [27, 28]. The first pair of parabolic energy dispersions nearest to E_F is seriously distorted along the $\pm\hat{k}_y$ directions, as shown for $\delta = 11b/8$ in figure 7.2(e). An eye-shaped stateless region is created with two distinct Dirac points at different energies; furthermore, the two neighboring conduction (valence) bands have strong hybridization. Finally, such energy bands are transformed into two pairs of isotropic Dirac cone structures at $\delta = 1.5b$ (figure 7.2(f)). In particular, the two Dirac cones are slightly tilted, and the Dirac points deviate from the K point (figure 7.3) [23]. The AA′ and AA stacking, respectively, possess non-vertical and vertical Dirac cone structures. This will induce critical differences in the magnetic quantization and magnetic-optical properties. The above-mentioned dramatic transformation of the energy bands is also confirmed by the first principle calculations [15].

The 2D electronic structures, as shown in figure 7.4(a)–(f), which are depicted along the high symmetry points, can also provide useful information on the electronic properties. The low-lying isotropic energy dispersions near E_F are transformed between the isotropic and anisotropic characteristics (insets). AA stacking shows linear isotropic dispersions near the K point (figure 7.4(a)). During the variation of the shift, they change into anisotropic distorted dispersions at $\delta = b/8$ and $6b/8$ (figure 7.4(b) and (c)) and then become parabolic isotropic dispersions at $\delta = b$ (figure 7.4(d)). Then, the energy dispersions are anisotropic in the range of $b \leqslant \delta \leqslant 1.5b$ (figure 7.4(e) and (f)). Except for the AA and AB stacking, other systems show distorted energy dispersions and highly anisotropic 2D band structures. Regardless of the semimetallic property of sliding BLG, tiny energy spacings can exist along K → M and K → M′ (figure 7.4(b) and (c)). Concerning the middle-energy electronic states, the M and M′ points remain as saddle- points in the energy wave vector space; that is, they will contribute greatly to the DOS and absorption spectrum, and their energies split, except for the AA and AB stacking. The three-fold rotation symmetry is destroyed, while the space and $y \rightarrow -y$ inversion symmetries are invariant. This accounts for the distinct energies between the two M points and the four M′ points in the first Brillouin zone. The splitting energy is observable with a wide range of ≈ 0.1– –0.6 eV, when the sliding BLG significantly deviates from AA or AB stacking.

Up to now, ARPES measurements have confirmed two vertical Dirac cones in AA stacking and two pairs of parabolic bands in AB stacking [27, 28]. The feature-rich electronic properties of configuration-dependent BLG, including the transformation between the Dirac cone structures and the parabolic bands, the tilted Dirac cones, the stateless arc-shaped region, the distorted energy dispersions and the

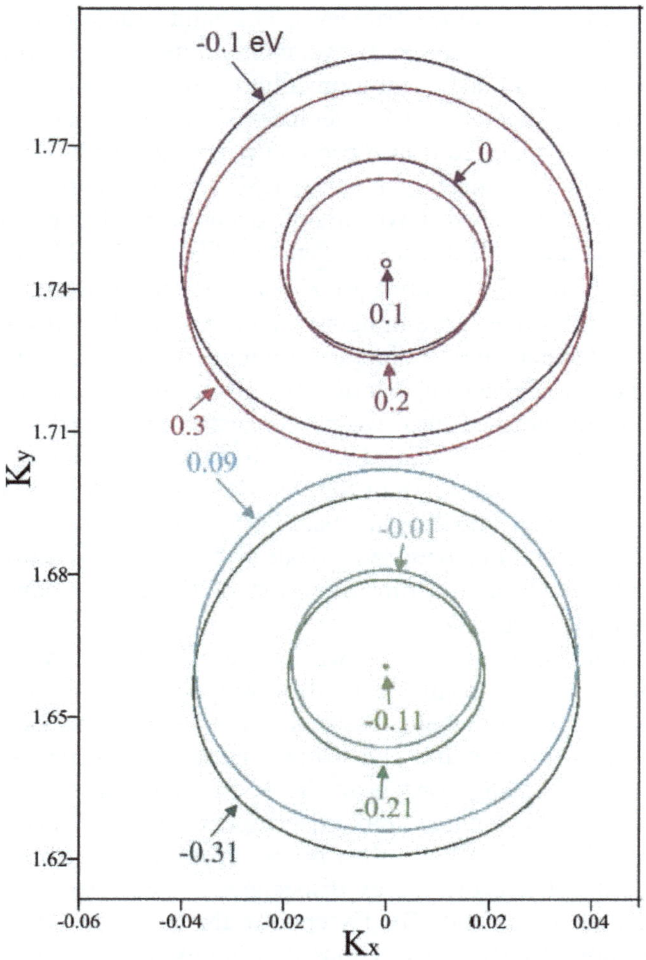

Figure 7.3. The constant energy contours of $\delta = 12b/8$ related to the two tilted Dirac cone structures.

splitting of the middle-energy states, can be further verified with ARPES [27–30]. This is one way to identify the tiny misalignment stacking related to various BLG configurations, as suggested by the experimental examinations on the predicted phonon spectra [31].

Unusual electronic structures can diversify the absorption spectra. The AA stacking reveals a forbidden excitation zone before the threshold frequency, since the spatial wavefunction symmetries lead to zero velocity matrix elements for the inter-Dirac-cone excitations. Two shoulders are presented in the absorption spectra, as indicated in figure 7.5(a). The frequencies of 0.64 and 0.73 eV are proportional to the Fermi velocities of the upper and lower Dirac cones, respectively; therefore, the optical measurements on the initial absorption structures can be used to identify the Dirac cone Fermi velocity. The optical gap vanishes in a small shift ($\delta = b/8$ in figure 7.5(b)), being replaced by a prominent peak at ~0.1 eV and a forbidden region with

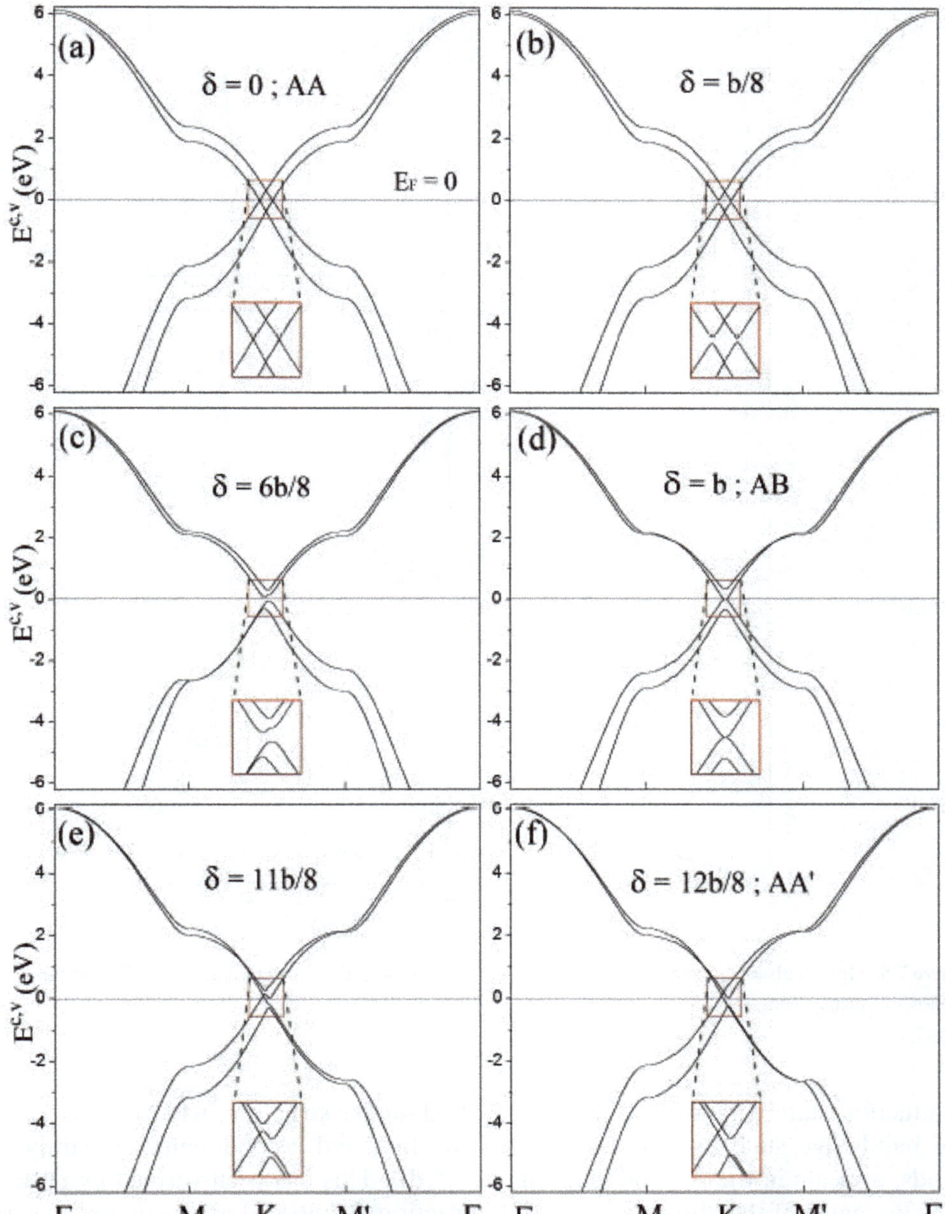

Figure 7.4. The 2D band structures of various types of stacking: (a) $\delta = 0$, (b) $b/8$, (c) $6b/8$, (d) b, (e) $11b/8$; (f) $12b/8$. Also shown in the insets are the low-lying ones.

a width of ~0.4 eV. Such features are associated with the saddle-points and the vacant arc-shaped region in the distorted energy bands (figure 7.4(b)). Specifically, the $\delta = 6b/8$ stacking exhibits a gapless and featureless absorption spectrum as a result of the quite different wave vectors in the saddle points of the valence and

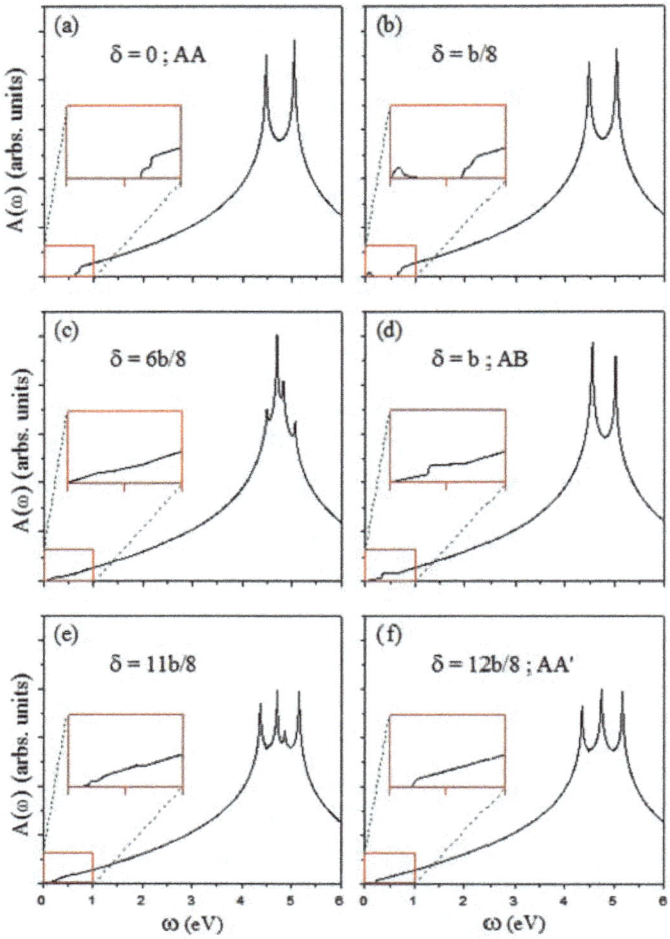

Figure 7.5. Optical absorption spectra of (a) $\delta = 0$, (b) $b/8$, (c) $6b/8$, (d) b, (e) $11b/8$, (f) $12b/8$. The insets show the low-frequency ones.

conduction states (figure 7.5(c)). A ~0.3 eV shoulder structure, which comes from the band-edge state excitations between the first and second pairs of parabolic bands, appears in the AB stacking (figure 7.5(d)). This has been verified by optical measurements of the absorption [32, 33], reflection [34–36, 38], transmission spectra [38–40] and Raman scattering [41, 42]. Moreover, an optical gap is recovered in the large-shift BLG, e.g. $\delta = 11b/8$ in figure 7.5(e) and $\delta = 12b/8$ in figure 7.5(f). As for the middle-frequency optical spectra, the AA, small-δ and AB stacking (figure 7.5(a), (b) and (d)), have two very strong absorption peaks, while the others present four or three (figure 7.5(c), (e) and (f)). This directly reflects the existence or destruction of the three-fold rotation symmetry. Obviously, experimentally measuring [41, 42, 46] the low- and middle-frequency absorption structures is an efficient way of identifying the various stacking configurations in sliding BLG.

7.2 Magneto-electronic and magneto-optical properties

Three types of LLs, well-behaved, perturbed and undefined, are present in the different stacking systems. The main characteristics are identified from the spatial distributions of the wavefunctions and the B_0-dependent energy spectra. Both the AA and AA′ stacking belong to the first type. The first and second types can be found in most stacking systems, such as AB, ABC and AAB, as well as sliding BLG with a small shift. The quantum numbers in the sliding systems are defined by the B^1 sublattice for the localization center of 2/6. Moreover, the third type is revealed in the specific stacking systems corresponding to the dramatic changes of the energy bands. In addition, the degeneracy of the LL states remains the same for any shift.

The LLs of AA and AA′ stacking exhibit regular spatial distributions with the symmetry or antisymmetry about the localization centers and the normal zero points. Their energies are proportional to $\sqrt{n^{c,v}B_0}$, as revealed in the B_0-dependent energy spectrum ($\delta = b$ in figure 7.7(d)). However, the important difference between them lies in the localization centers. AA stacking has a fixed localization center of 2/6 (figure 7.6(a)-(d)(a)). As for AA′ stacking, the first and second groups of LL wavefunctions deviate from the 2/6 position in opposite directions ($n_1^{c,v}$ and $n_2^{c,v}$ in figure 7.6(a)–(d)(f)). The main reason for this is that the k_y component is different for the current Dirac points and the K points, and the deviation of the Dirac cone centers grows with increasing energy (figure 7.3) [23]. The tilted Dirac cone structures can create new magneto-optical selection rules.

AB-stacked BLG has a high-symmetry configuration, while its Hamiltonian cannot be exactly diagonalized [47]. The interlayer atomic interactions in the magnetic quantizations can be treated as perturbations. The LL wavefunctions are presented with the normal quantum numbers, as shown in figure 7.6(a)-(d)(d) at $B_0 = 40$ T. However, when the magnetic field is very high ($B_0 \gtrsim 60$ T; in the inset of figure 7.7(c)), the B_0-dependent electronic spectra show that some perturbed LLs in the anticrossing regions are related to distinct groups. The second type of LL is frequently revealed in the sliding BLG with a small shift measured from the AA, AA′ and AB stacking, as indicated by the anticrossings in figure 7.6(a)-(d)(a). The low-lying LLs of $\delta = b/8$ exhibit distorted wavefunctions composed of the main and side modes (figure 7.6(a)–(d)(b)). Such LLs come from the low-energy distorted electronic structure (figure 7.2(b)).

With a large shift deviating from the high-symmetry stacking configurations, the LL wavefunctions present irregular oscillations without the normal zero points, e.g. $\delta = 6b/8$ and $11b/8$ in figure 7.6(a)–(d)(c) and (e), respectively. They cannot be defined as a single-mode oscillator; that is, the main mode is absent for any magnetic field strength. Moreover, the B_0-dependent energy spectrum indicates that each LL in the second group displays serious anticrossings with all the LLs in the first one, and vice versa (figure 7.7(b)). This clearly illustrates that such LLs consist of various zero-point modes, or that they could be verified as undefined LLs; they arise from the fully distorted energy bands (figure 7.2(c)). The K and K′ points do not belong to the critical points; therefore, the low-energy expansion about them is not suitable for magnetic quantization. It is impossible to comprehend the undefined modes from the effective mass approximation. The above-mentioned three types of LL have distinct

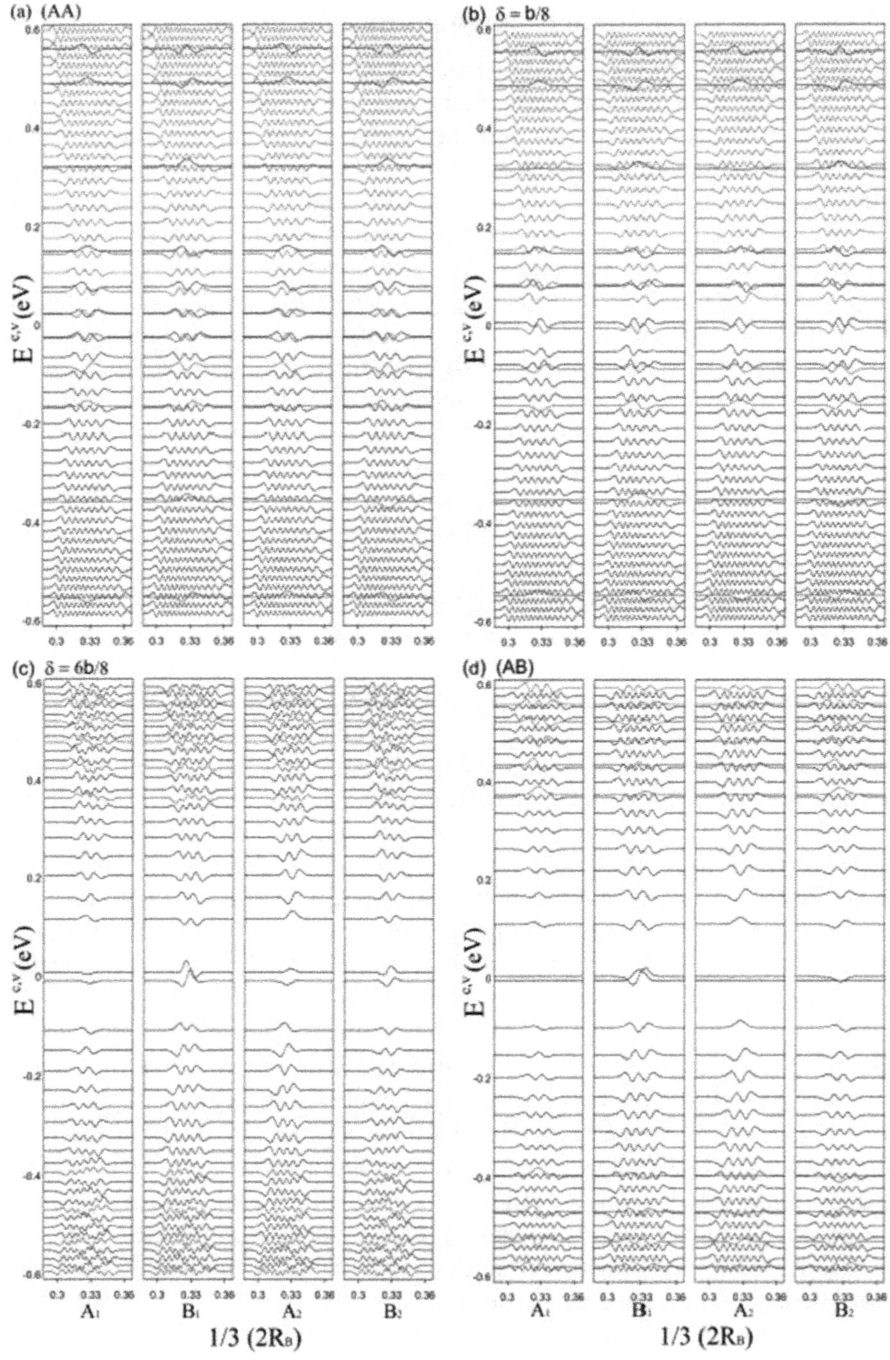

Figure 7.6. The LL energies and wavefunctions at $B_0 = 40$ T for (a) $\delta = 0$, (b) $b/8$, (c) $6b/8$, (d) b, (e) $11b/8$, (f) $12b/8$.

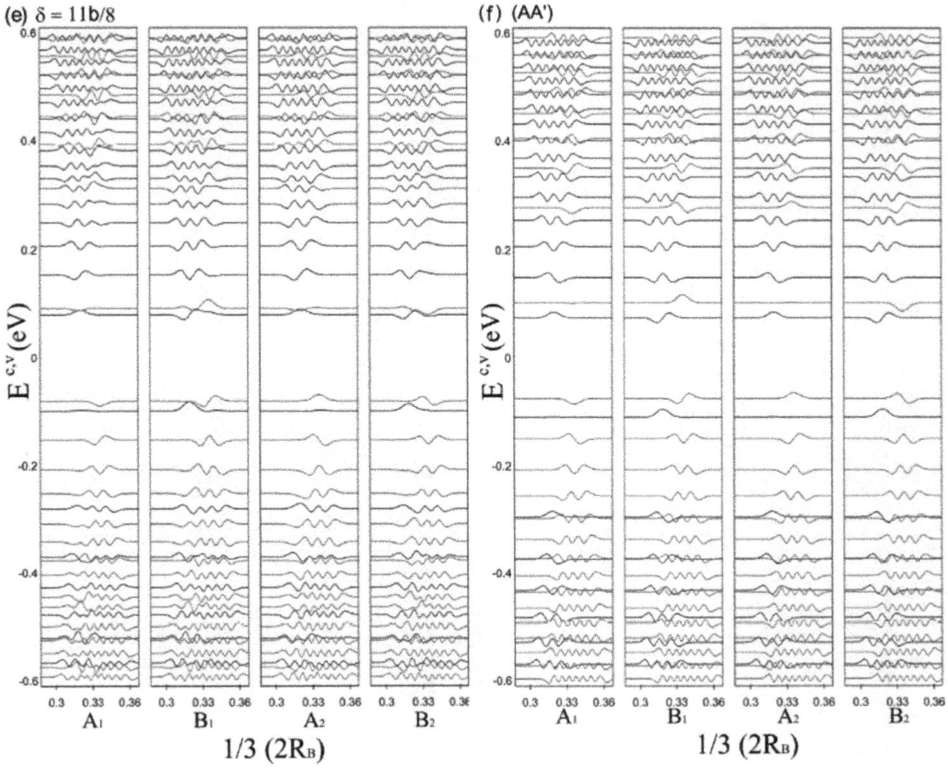

Figure 7.6. (Continued.)

energy spectra, in which their main characteristics can be directly examined by STS measurements [48–51].

The diverse spatial distributions of three kinds of LL can enrich the magneto-optical absorption spectra [23]. The absorption spectra of sliding BLG are revealed in the twin- or single-peak structures, depending on the difference of excitation frequency in the asymmetric LL energies. The well-behaved LLs in the AA and AA′ stacking exhibit two categories of intragroup excitation ($n_1^c \rightarrow n_1^v$ and $n_2^c \rightarrow n_2^v$). The absorption peaks of the former satisfy a specific selection rule of $\Delta n = \pm 1$ (figure 7.8(a)), while those of the latter are described by $\Delta n = \pm 1$, 0 and $\pm$ 2 (figure 7.8(f)). The new selection rules of $\Delta n = 0$ and ± 2, which correspond to a lot of observable absorption peaks (blue and green triangles), are associated with the tilted Dirac cone structures in the AA′ stacking. They are mainly determined by the relations between the initial state of the A^l (B^l) sublattice and the final state of the B^l (A^l) sublattice. The valence and conduction state subenvelope functions have different localization centers, so that the excitation channels of $\Delta n = 0$ and $\pm$ 2 gradually become obvious at higher frequencies. Such rules are not predicted to exist in other condensed matter systems with well-behaved LLs.

On the other hand, four categories of inter-LL excitations, two intragroup and two intergroup, are revealed in the other stacking systems. AB stacking, with the

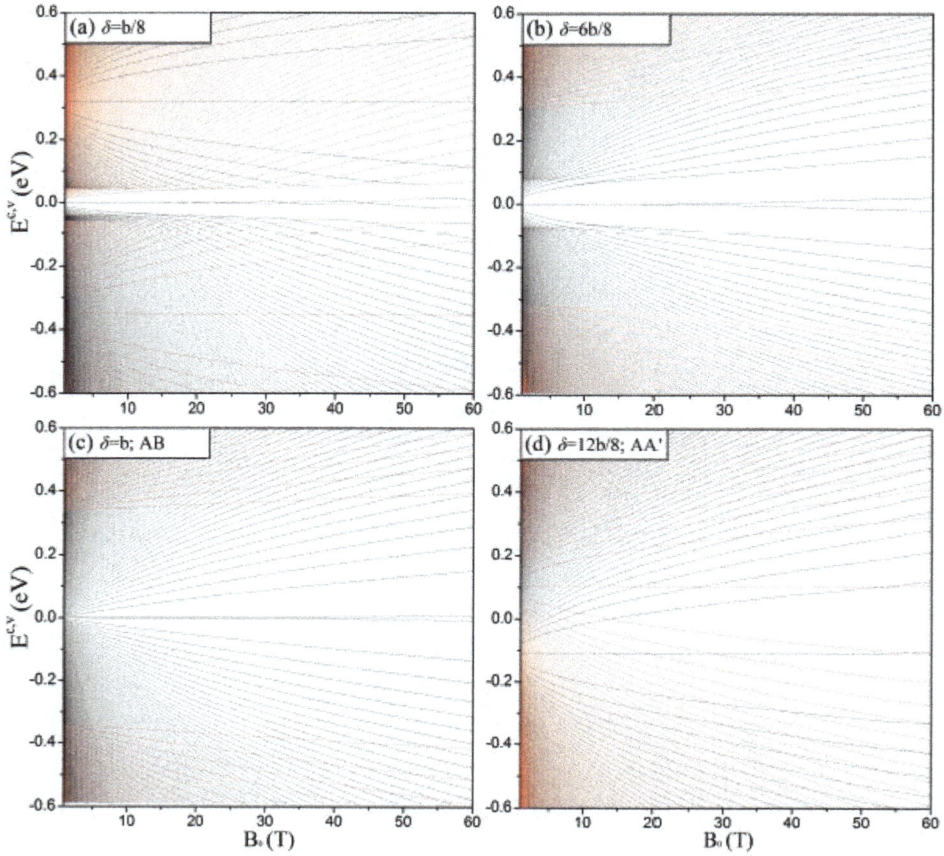

Figure 7.7. The B_0-dependent LL energy spectra of (a) $b/8$, (b) $6b/8$, (c) b, (d) $12b/8$.

well-behaved LLs at $B_0 \lesssim 60$ T, exhibits many prominent absorption peaks of $\Delta n = \pm 1$, as indicated in figure 7.8(d). Specifically, the absorption frequencies due to the inter-LL excitations of the first group present the linear B_0-dependence (which is identified in figure 7.7(c)). As for the sliding BLG, which slightly deviates from the normal stacking, most of the absorption peaks agree with $\Delta n = \pm 1$ ($\delta = b/8$ in figure 7.8(b)). However, some peaks have the extra selection rules of $\Delta n = 0$ and $\pm$ 2 (blue and green triangles), whose intensities are dominated by the relative strength of the main and side modes in the perturbed LLs (figure 7.6(a)-(d)(b)). In particular, the undefined LLs show a large number of absorption peaks without any optical selection rules. This means that all the occupied LLs can be effectively excited by any unoccupied ones, i.e. all the inter-LL excitation channels are effective under the requirement of the Fermi–Dirac distribution. In general, such peaks are relatively low for easy observation, since the velocity matrix elements are significantly reduced by the random spatial distributions of the disordered quantum modes. The above-mentioned unique magneto-optical properties, the new selection rules, the extra ones and the absence of a selection rule, are worthy of a detailed examination using optical spectroscopy [34, 39, 40, 52].

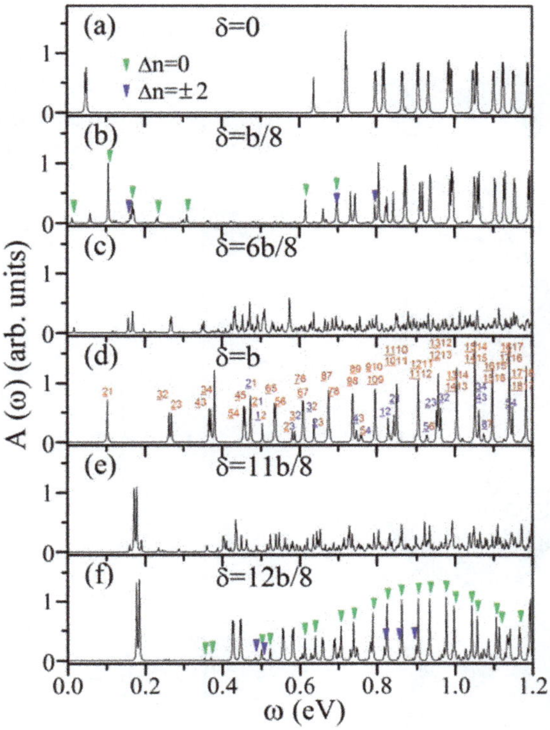

Figure 7.8. Magneto-absorption spectra at $B_0 = 40$ T for (a) $\delta = 0$, (b) $b/8$, (c) $6b/8$, (d) b, (e) $11b/8$, (f) $12b/8$.

7.3 Electric-field-enriched optical properties

Electronic structures are significantly altered by the perpendicular electric field, such as Dirac cone structures, saddle-points, band gaps and energy dispersions. All the stacking systems have $y \to -y$ symmetry (and time reversal symmetry); that is, the electronic states are identical in the K and K′ valleys. Only AA stacking presents similar isotropic Dirac cone structures [53], since the $\mathbf{r} \to -\mathbf{r}$ and $y \to -y$ inversion symmetries remain the same. However, for AA′ stacking (figure 7.9(e)), the electric potential difference between two layers induces the splitting of the Dirac points (four points) and the seriously distorted Dirac cone structure with extra saddle-points. Energy gaps are generated in most of the bilayer stacking systems except for those close to AA and AA′, as shown in figure 7.9(a) and (c) ($\delta = 6b/8$ and b). Specifically, AB stacking possesses weakly dispersive energy bands with a deep pocket near the Fermi level (the slightly distorted sombrero-shaped bands). The Dirac cone structures and energy gaps are directly reflected in the optical gaps. The electric field causes each stacking system to exhibit a threshold frequency (figure 7.9(b), (d) and (f)), which is rather different from the zero-field case. The dependence of the optical gap on the field strength is monotonous, i.e. it gradually grows as in figure 7.9(b) and (d) and declines as in figure 7.9(f) with the increasing field strength. Moreover, there exist extra special structures in the absorption spectrum.

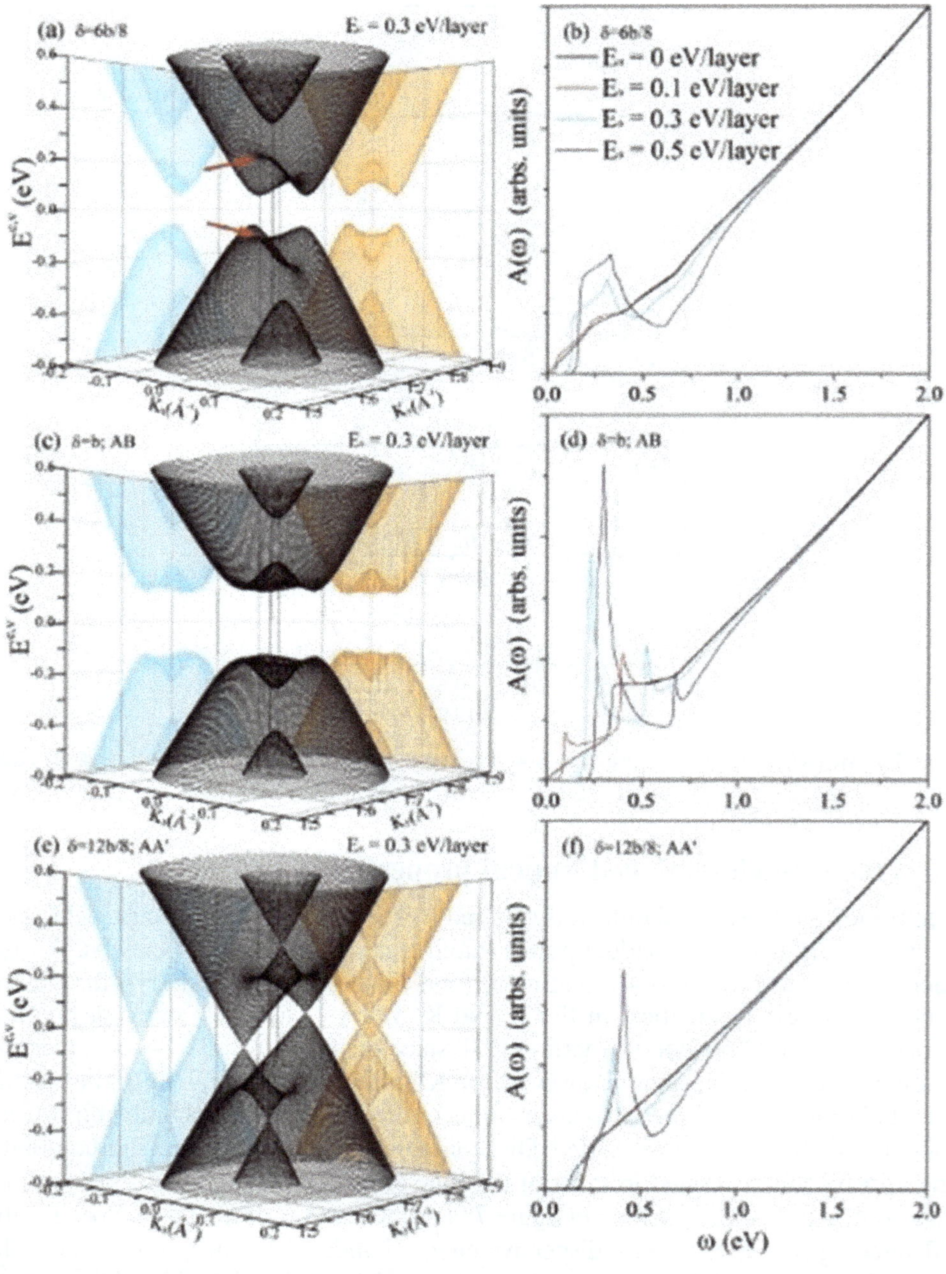

Figure 7.9. The low-energy electronic structures and optical spectra at distinct electric fields for bilayer graphene with (a) and (b) $\delta = 6b/8$, (c) and (d) b, (e) and (f) $12b/8$.

For the $\delta = 6b/8$ stacking, an extra absorption peak appears at a higher frequency (~0.35 eV), mainly owing to the nearly identical wave vectors in the valence and conduction saddle-points (red arrows). AB stacking has two prominent asymmetric peaks, in which those of a lower- and higher-frequency arise from the optical excitations of the intra- and interpair energy bands, respectively. The high DOS from the first pair of energy bands accounts for the strong absorption intensity. The **E**-induced saddle points in the AA′ stacking can create quite a strong peak structure.

In the presence of a composite electric and magnetic field, the $x \to -x$ and $y \to -y$ inversion symmetries are conserved in the AA and AA′ stacking, while the former is absent in the other BLG stacking. Under the condition of splitting, the magnetic quantization of the K and K′ valleys indicates that only these two types of high-symmetry stacking possess four-fold degenerate LLs. Splitting LLs exist in the other types of BLG. Moreover, the seriously distorted and weakly dispersive energy dispersions are magnetically quantized in multimode LLs. There are more perturbed and undefined LLs, so that LL anticrossings occur frequently in the E_0- and B_0-dependent energy spectra. Apparently, the destruction of the state degeneracy and the well-behaved mode induce more complicated magneto-absorption structures, as shown in figure 7.10(a)–(c). The number of absorption peaks is largely enhanced. Most of them present a lower intensity, except those whose LLs are associated with the high-DOS energy dispersions (0.2 eV $\leqslant \omega \leqslant$ 0.4 eV). The original magneto-selection rules can survive in a composite field.

Optical measurements under external fields are only taken for the AB-stacked BLG using absorption [32, 33], transmission [38–40] reflection [34–36, 38] and Raman scattering [7, 41, 42] spectroscopies. They agree with a number of theoretical predictions including the low-frequency shoulder structure at zero fields [22, 43–45, 54], the **E**-induced threshold frequency and two prominent peaks [32–33], the magneto-optical selection rule of $\Delta n = \pm 1$ for the inter-LL excitations of the first group [34], and the linearly B_0-dependent absorption frequency. The other predicted results, which are the forbidden excitation region induced by the Fermi–Dirac distribution, the **E**-created optical gaps and special structures in any bilayer stacking, the diverse selection rules due to three kinds of LL, and the drastic changes of the spectral features in a composite field, can be further examined by optical spectroscopy.

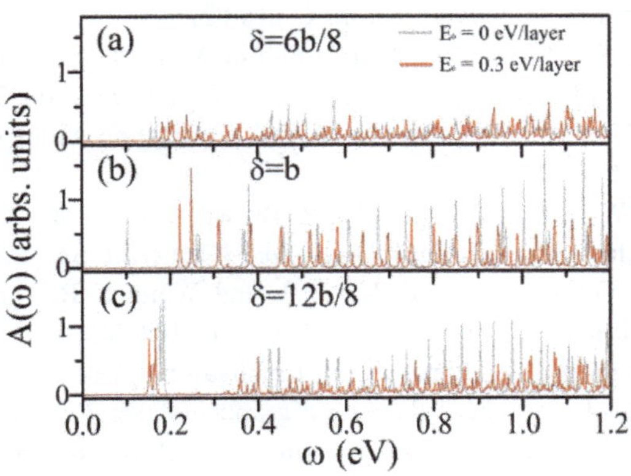

Figure 7.10. Absorption spectra under a composite field for bilayer graphene with (a) $\delta = 6b/8$, (b) b, (c) $12b/8$.

References

[1] Daboussi A, Mandhour L, Fuchs J N and Jaziri S 2014 Tunable zero-energy transmission resonances in shifted graphene bilayer *Phys. Rev.* B **89** 085426

[2] Kim K S, Walter A L, Moreschini L, Seyller T, Horn K and Rotenberg E *et al* 2013 Coexisting massive and massless Dirac fermions in symmetry-broken bilayer graphene *Nat. Mater.* **12** 887–92

[3] Fang W, Hsu A L, Caudillo R, Song Y, Birdwell A G and Zakar E *et al* 2013 Rapid identification of stacking orientation in isotopically labeled chemical-vapor grown bilayer graphene by Raman spectroscopy *Nano Lett.* **13** 1541–8

[4] Popov A M, Lebedeva I V, Knizhnik A A, Lozovik Y E and Potapkin B V 2011 Commensurate-incommensurate phase transition in bilayer graphene *Phys. Rev.* B **84** 045404

[5] Perebeinos V, Tersoff J and Avouris P 2012 Phonon-mediated interlayer conductance in twisted graphene bilayers *Phys. Rev. Lett.* **109** 236604

[6] Cocemasov A I, Nika D L and Balandin A A 2013 Phonons in twisted bilayer graphene *Phys. Rev.* B **88** 035428

[7] Poncharal P, Ayari A, Michel T and Sauvajol J L 2008 Raman spectra of misoriented bilayer graphene *Phys. Rev.* B **78** 113407

[8] Zheng J, Guo P, Ren Z, Jiang Z, Bai J and Zhang Z 2012 Conductance fluctuations as a function of sliding motion in bilayer graphene nanoribbon junction: a first-principles investigation *Appl. Phys. Lett.* **101** 083101

[9] Son Y W, Choi S M, Hong Y P, Woo S and Jhi S H 2011 Electronic topological transition in sliding bilayer graphene *Phys. Rev.* B **84** 155410

[10] Lee K W and Lee C E 2015 Extreme sensitivity of the electric field-induced band gap to the electronic topological transition in sliding bilayer graphene *Sci. Rep.* **5** 17490

[11] Lopes dos Santos J M B, Peres N M R and Castro Neto A H 2007 Graphene bilayer with a twist: electronic structure *Phys. Rev. Lett.* **99** 256802

[12] Bhattacharyya S and Singh A K 2016 Lifshitz transition and modulation of electronic and transport properties of bilayer graphene by sliding and applied normal compressive strain *Carbon* **99** 432–8

[13] Lei Y, Jiang W, Dai X, Song R, Wang B and Gao Y *et al* 2015 Slippage in stacking of graphene nanofragments induced by spin polarization *Sci. Rep.* **5** 10985

[14] Wong J H, Wu B R and Lin M F 2012 Strain effect on the electronic properties of single layer and bilayer graphene *J. Phys. Chem.* C **116** 8271–7

[15] Tran N T T, Lin S Y, Glukhova O E and Lin M F 2015 Configuration-induced rich electronic properties of bilayer graphene *J Phys. Chem.* C **119** 10623–30

[16] Feng X, Kwon S, Park J Y and Salmeron M 2013 Superlubric sliding of graphene nanoflakes on graphene *ACS Nano* **7** 1718–24

[17] Alden J S, Tsen A W, Huang P Y, Hovden R, Brown L and Park J *et al* 2013 Strain solitons and topological defects in bilayer graphene *Proc. Natl Acad. Sci. USA* **110** 11256–60

[18] Liu Z, Yang J R, Grey F, Liu J Z, Liu Y L and Wang Y B *et al* 2012 Observation of microscale superlubricity in graphite *Phys. Rev. Lett.* **108** 205503

[19] Lebedeva I V, Knizhnik A A, Popov A M, Ershova O V, Lozovik Y E and Potapkin B V 2010 Fast diffusion of a graphene flake on a graphene layer *Phys. Rev.* B **82** 155460

[20] Koshino M 2013 Electronic transmission through AB–BA domain boundary in bilayer graphene *Phys. Rev.* B **88** 115409

[21] Nemec N and Cuniberti G 2007 Hofstadter butterflies of bilayer graphene *Phys. Rev.* B **75** 201404

[22] Chen S C, Chiu C W, Wu C L and Lin M F 2014 Shift-enriched optical properties in bilayer graphene *RSC Adv.* **4** 63779–83

[23] Huang Y K, Chen S C, Ho Y H, Lin C Y and Lin M F 2014 Feature-rich magnetic quantization in sliding bilayer graphenes *Sci. Rep.* **4** 7509

[24] Saito R, Dresselhaus G and Dresselhaus M S 1993 Electronic structure of double layer graphene tubules *J. Appl. Phys.* **73** 494–500

[25] Charlier J C and Michenaud J P 1993 Energetics of multilayered carbon tubules *Phys. Rev. Lett.* **70** 1858–61

[26] Koshino M and McCann E 2010 Parity and valley degeneracy in multilayer graphene *Phys. Rev.* B **81** 115315

[27] Mucha-Kruczynski M, Tsyplyatyev O, Grishin A, McCann E, Fal'ko V I and Bostwick A *et al* 2008 Characterization of graphene through anisotropy of constant-energy maps in angle-resolved photoemission *Phys. Rev.* B **77** 195403

[28] Moreau E, Godey S, Wallart X, Razado-Colambo I, Avila J and Asensio M C *et al* 2013 High-resolution angle-resolved photoemission spectroscopy study of monolayer and bilayer graphene on the C-face of SiC *Phys. Rev.* B **88** 075406

[29] Liu Y, Zhang L, Brinkley M K, Bian G, Miller T and Chiang T C 2010 Phonon-induced gaps in graphene and graphite observed by angle-resolved photoemission *Phys. Rev. Lett.* **105** 136804

[30] Ayria P, Nugraha A R T, Hasdeo E H, Czank T R, Tanaka S and Saito R 2015 Photon energy dependence of angle-resolved photoemission spectroscopy in graphene *Phys. Rev.* B **92** 195148

[31] Choi S M, Jhi S H and Son Y W 2015 Anomalous optical phonon splittings in sliding bilayer graphene *ACS nano* **7** 7151–6

[32] Geng B, Horng J, Zhang Y, Tang T T, Park C H and Girit C *et al* 2010 Optical spectroscopy of bilayer graphene *Phys. Status Solidi* B **247** 2931–4

[33] Zhang Y, Tang T T, Girit C, Hao Z, Martin M C and Zettl A *et al* 2009 Direct observation of a widely tunable bandgap in bilayer graphene *Nature* **459** 820–3

[34] Berciaud S, Potemski M and Faugeras C 2014 Probing electronic excitations in mono- to pentalayer graphene by micro magneto-Raman spectroscopy *Nano Lett.* **14** 4548–53

[35] Kuzmenko A B, Crassee I, van der Marel D, Blake P and Novoselov K S 2009 Determination of the gate-tunable band gap and tight-binding parameters in bilayer graphene using infrared spectroscopy *Phys. Rev.* B **80** 165406

[36] Kuzmenko A B, van Heumen E, van der Marel D, Lerch P, Blake P and Novoselov K S *et al* 2009 Infrared spectroscopy of electronic bands in bilayer graphene *Phys. Rev.* B **79** 115441

[38] Zhang L M, Li Z Q, Basov D N, Fogler M M, Hao Z and Martin M C 2008 Determination of the electronic structure of bilayer graphene from infrared spectroscopy *Phys. Rev.* B **78** 235408

[38] Zhang L M, Li Z Q, Basov D N, Fogler M M, Hao Z and Martin M C 2008 Determination of the electronic structure of bilayer graphene from infrared spectroscopy *Phys. Rev.* B **78** 235408

[39] Faugeras C, Berciaud S, Leszczynski P, Henni Y, Nogajewski K and Orlita M *et al* 2015 Landau level spectroscopy of electron–electron interactions in graphene *Phys. Rev. Lett.* **114** 126804

[40] Orlita M, Faugeras C, Barra A L, Martinez G, Potemski M and Basko D M *et al* 2015 Infrared magneto-spectroscopy of two-dimensional and three-dimensional massless fermions: a comparison *J. Appl. Phys.* **117** 112803

[41] Malard L M, Nilsson J, Elias D C, Brant J C, Plentz F and Alves E S *et al* 2007 Probing the electronic structure of bilayer graphene by Raman scattering *Phys. Rev.* B **76** 201401

[42] Ferrari A C, Meyer J C, Scardaci V, Casiraghi C, Lazzeri M and Mauri F *et al* 2006 Raman spectrum of graphene and graphene layers *Phys. Rev. Lett.* **97** 187401

[43] Lu C L, Chang C P, Huang Y C, Chen R B and Lin M L 2006 Influence of an electric field on the optical properties of few-layer graphene with AB stacking *Phys. Rev.* B **73** 144427

[44] Maka K F, Sfeirb M Y, Misewichb J A and Heinza T F 2010 The evolution of electronic structure in few-layer graphene revealed by optical spectroscopy *Proc. Natl Acad. Sci. USA* **107** 14999–15004

[45] Kim D S, Kwon H, Nikitin A Y, Ahn S, Martin-Moreno L and Garcia-Vidal F J *et al* 2015 Stacking structures of few-layer graphene revealed by phase-sensitive infrared nanoscopy *ACS Nano* **9** 6765–73

[46] Liou S C, Shie C S, Chen C H, Breitwieser R, Pai W W and Guo G Y *et al* 2015 π-plasmon dispersion in free-standing graphene by momentum-resolved electron energy-loss spectroscopy *Phys. Rev.* B **91** 045418

[47] McCann E and Fal'ko V I 2006 Landau-level degeneracy and quantum Hall effect in a graphite bilayer *Phys. Rev. Lett.* **96** 086805

[48] Matsui T, Kambara H, Niimi Y, Tagami K, Tsukada M and Fukuyama H 2005 STS observations of Landau levels at graphite surfaces *Phys. Rev.* B **94** 226403

[49] Uchoa B, Yang L, Tsai S W, Peres N M and Rm Castro Neto A H 2009 Theory of scanning tunneling spectroscopy of magnetic adatoms in graphene *Phys. Rev. Lett.* **103** 206804

[50] Chae J, Jung S, Young A F, Dean C R, Wang L and Gao Y *et al* 2012 Renormalization of the graphene dispersion velocity determined from scanning tunneling spectroscopy *Phys. Rev. Lett.* **109** 116802

[51] Li G, Luican A and Andrei E Y 2009 Scanning tunneling spectroscopy of graphene on graphite *Phys. Rev. Lett.* **102** 176804

[52] Faugeras C, Binder J, Nicolet A A L, Leszczynski P, Kossacki P and Wysmolek A *et al* 2014 A micro-magneto-Raman scattering study of graphene on a bulk graphite substrate *EPL* **108** 27011

[53] Tsai S J, Chiu Y H, Ho Y H and Lin M F 2012 Gate-voltage-dependent Landau levels in AA-stacked bilayer graphene *Chem. Phys. Lett.* **550** 104–110

[54] Mak K F, Lui C H, Shan J and Heinz T F 2009 Observation of an electric-field-induced band gap in bilayer graphene by infrared spectroscopy *Phys. Rev. Lett.* **102** 256405

[55] Li X L, Wang X R, Zhang L, Lee S W and Dai H J 2008 Chemically derived, ultrasmooth graphene nanoribbon semiconductors *Science* **319** 1229–32

[56] Fujii S and Enoki T 2010 Cutting of oxidized graphene into nanosized pieces *J. Am. Chem. Soc.* **132** 10034–41

[57] Kosynkin D V, Higginbotham A L, Sinitskii A, Lomeda J R, Dimiev A and Price B K *et al* 2009 Longitudinal unzipping of carbon nanotubes to form graphene nanoribbons *Nature* **458** 872–U5

[58] Jiao L, Zhang L, Wang X, Diankov G and Dai H J 2009 Narrow graphene nanoribbons from carbon nanotubes *Nature* **458** 877–80

[59] Kim K, Sussman A and Zettl 2010 A Graphene nanoribbons obtained by electrically unwrapping carbon nanotubes *ACS Nano* **4** 1362–6

[60] Blankenburg S, Cai J, Ruffieux P, Jaafar R, Passerone D and Feng X *et al* 2012 Intraribbon heterojunction formation in ultranarrow graphene nanoribbons *ACS Nano* **6** 2020–5
[61] Yi Z, Yanfang Z, Geng L, Jianchen L, Xiao L and Shixuan D *et al* 2014 Direct visualization of atomically precise nitrogen-doped graphene nanoribbons *Appl. Phys. Lett.* **105** 023101
[62] Xie L M, Wang H L, Jin C H, Wang X R, Jiao L Y and Suenaga K *et al* 2011 Graphene nanoribbons from unzipped carbon nanotubes: atomic structures, Raman spectroscopy, and electrical properties *J. Am. Chem. Soc.* **133** 10394–7
[63] Chang S L, Wu B R, Wong J H and Lin M F 2014 Configuration-dependent geometric and electronic properties of bilayer graphene nanoribbons *Carbon* **77** 1031–9
[64] Lin M F and Shyu F L 2000 Optical properties of nanographite ribbons *J. Phys. Soc. Jpn* **69** 3529–32
[65] Hsu H and Reichl L E 2007 Selection rule for the optical absorption of graphene nanoribbons *Phys. Rev.* B **76** 045418
[66] Chung H C, Lee M H, Chang C P and Lin M F 2011 Exploration of edge-dependent optical selection rules for graphene nanoribbons *Opt. Express* **19** 23350–63
[67] Sasaki K I, Kato K, Tokura Y, Oguri K and Sogawa T 2010 Theory of optical transitions in graphene nanoribbons *Phys. Rev.* B **84** 085458
[68] Huang Y C, Chang C P and Lin M F 2007 Magnetic and quantum confinement effects on electronic and optical properties of graphene ribbons *Nanotechnology* **18** 495401
[69] Chang C P, Huang Y C, Lu C L, Ho J H, Li T S and Lin M F 2006 Electronic and optical properties of a nanographite ribbon in an electric field *Carbon* **44** 508–15
[70] Chung H C, Chang C P, Lin C Y and Lin M F 2016 Electronic and optical properties of graphene nanoribbons in external fields *Phys. Chem. Chem. Phys.* **18** 7573–616
[71] Soule D E 1958 Magnetic field dependence of the Hall effect and magnetoresistance in graphite single crystals *Phys. Rev* **112** 698–707
[72] Galt J K, Yager W A and Dail H W 1956 Cyclotron resonance effects in graphite *Phys. Rev* **103** 1586–7
[73] Dresselhaus M S and Dresselhaus G 1981 Intercalation compounds of graphite *Adv. Phys* **30** 139–326
[74] Chiu C W, Lee S H, Chen S C, Shyu F L and Lin M F 2010 Absorption spectra of AA-stacked graphite *New J. Phys.* **12** 083060
[75] Chiu C W, Huang Y C, Chen S C, Lin M F and Shyu F L 2011 Low-frequency electronic and optical properties of rhombohedral graphite *Phys. Chem. Chem. Phys.* **13** 6036–42
[76] Chen R B, Chiu Y H and Lin M F 2013 A theoretical evaluation of the magneto-optical properties of AA-stacked graphite *Carbon* **54** 268–76
[77] Chen R B, Chiu Y H and Lin M F 2015 Beating oscillations of magneto-optical spectra in simple hexagonal graphite *Comput. Phys. Commun.* **189** 60–5
[78] Nakao K 1976 Landau level structure and magnetic breakthrough in graphite *J. Phys. Soc. Jpn.* **40** 761–8
[79] Ho Y H, Wang J, Chiu Y H, Lin M F and Su W P 2011 Characterization of Landau subbands in graphite: a tight-binding study *Phys. Rev.* B **83** 121201
[80] Ho C H, Chang C P and Lin M F 2013 Precessing anisotropic Dirac cone and Landau subbands along a nodal spiral *New J. Phys.* **15** 053032
[81] Ho C H, Chang C P and Lin M F 2015 Optical magnetoplasmons in rhombohedral graphite with a three-dimensional dirac cone structure *J. Phys. Condens. Matter* **27** 125602

[82] Aufray B, Kara A, Vizzini S, Oughaddou H, Leandri C and Ealet B *et al* 2010 Graphene-like silicon nanoribbons on Ag(110): a possible formation of silicene *Appl. Phys. Lett.* **96** 183102
[83] Lalmi B, Oughaddou H, Enriquez H, Kara A, Vizzini S and Ealet B *et al* 2010 Epitaxial growth of a silicene sheet *Appl. Phys. Lett.* **97** 223109
[84] Vogt P, de Padova P, Quaresima C, Avila J, Frantzeskakis E and Asensio M C *et al* 2012 Silicene: compelling experimental evidence for graphenelike two-dimensional silicon *Phys. Rev. Lett.* **108** 155501
[85] Bianco E, Butler S, Jiang S, Restrepo O, Windl W and Goldberger J 2013 Stability and exfoliation of germanane: a germanium graphane analogue *ACS Nano* **7** 4414–21
[86] Chu J H, Kwak J, Kim S D, Lee M J, Kim J J and Park S D *et al* 2014 Monolithic graphene oxide sheets with controllable composition *Nature Comm.* **5** 3383
[87] Derivaz M, Dentel D, Stephan R, Hanf M C, Mehdaoui A and Sonnet P *et al* 2015 Continuous germanene layer on A(111) *Nano Lett.* **15** 2510–6
[88] Zhu F F, Chen W J, Xu Y, Gao C L, Guan D D and Liu C H *et al* 2015 Epitaxial growth of two-dimensional stanene *Nat. Mater.* **14** 1020
[89] Sun Z, Hasan T, Torrisi F, Popa D, Privitera G and Wang F *et al* 2010 Graphene mode-locked ultrafast laser *ACS Nano* **4** 803–10
[90] Tao L, Cinquanta E, Chiappe D, Grazianetti C, Fanciulli M and Dubey M *et al* 2015 Silicene field-effect transistors operating at room temperature *Nat. Nanotech.* **10** 227–31
[91] Tsai W F, Huang C Y, Chang T R, Lin H, Jeng H T and Bansil A 2013 Gated silicene as a tunable source of nearly 100 percent spin-polarized electrons *Nat. Commun.* **4** 1500
[92] Chen S C, Wu C L, Wu J Y and Lin M F 2016 Novel magnetic quantization of sp^3 bonding in monolayer tinene *Phys. Rev.* B **94** 045410
[93] Chang C P, Lu C L, Shyu F L, Chen R B, Huang Y C and Lin M F 2005 Magnetoelectronic properties of AB-stacked graphite *Carbon* **43** 1424–31
[94] Lu C L, Chang C P and Lin M F 2007 Magneto-electronic properties of AA- and ABC-stacked graphites *Eur. Phys. J.* B **60** 161–9
[95] Shyu F L and Lin M F 2001 Low-frequencyπ-electronic excitations of simple hexagonal graphite *J. Phys. Soc. Jpn* **70** 897–901
[96] Lin M F, Huang C S and Chuu D S 1997 Plasmons in graphite and stage-1 graphite intercalation compounds *Phys. Rev.* B **55** 13961
[97] Lin M F, Chuu D S, Huang C S, Lin Y K and Shung K W K 1996 Collective excitations in a single-layer carbon nanotube *Phys. Rev.* B **53** 15493
[98] Lin M F, Shyu F L and Chen R B 2000 Optical properties of well-aligned multiwalled carbon nanotube bundles *Phys. Rev.* B **61** 14114
[99] Watzel J, Pavlyukh Y, Schaffer A and Berakdar J 2016 Optical vortex driven charge current loop and optomagnetism in fullerenes *Carbon* **99** 439–3

Chiun-Yan Lin, Thi-Nga Do, Yao-Kung Huang and Ming-Fa Lin

Chapter 8

Concluding remarks

The generalized tight-binding mode and the gradient approximation are used to explore the rich absorption spectra of graphene. They are suitable for studying low-symmetry stacking configurations and complex external fields. The rich features due to the stacking configurations and the external fields are investigated in detail. AAA, ABA, ABC and AAB stacking, and the various types of stacking in BLG, are taken into account, and the zero, electric, magnetic and composite fields are included in the calculations. The presented results are compared with those from other theoretical calculations and validated by the experimental measurements, while most of the predictions require further experimental verification. The developed theoretical framework will promote future studies on other 2D materials [1–13].

High-symmetry AA-stacked trilayer graphene presents three vertical Dirac–Dirac structures with separate Dirac points. Only the intra-Dirac-cone optical excitations can survive because of the linearly symmetric or antisymmetric superposition of the tight-binding functions in the different layers. The spectral intensity due to the middle Dirac cone linearly depends on the excitation frequency, as observed in monolayer graphene. In addition, even-layer AA stacking possesses an optical gap determined by the first shoulder structure. The almost symmetric upper and lower Dirac cones, with free holes and electrons respectively, result in a forbidden excitation region under the specific fermion distributions. Their Dirac points shift away from the Fermi level in the presence of an electric field and thus create an enlarged vacant region or a blue shift in the shoulder structure. On the other hand, three obvious symmetric peaks, which come from the saddle-points near the M point, are revealed in the middle-frequency absorption spectrum. Three groups of well-behaved LLs only exhibit three categories of intragroup optical excitations, in which each prominent delta-function-like peak is characterized by the $\Delta n = \pm 1$ selection rule. Most spectral structures belong to single peaks with an almost uniform intensity, reflecting the approximately symmetric LL energy spectrum and the close Fermi velocities. Any peak frequency is characterized by the

doi:10.1088/978-0-7503-1566-1ch8 © IOP Publishing Ltd 2017

$\sqrt{B_0}$ dependence. As for the bilayer-like intragroup excitations, the first inter-LL channel changes discontinuously at certain magnetic fields, so that its frequency has a discontinuous B_0 dependence. In particular, the main features of the optical and electronic properties remain similar in a composite external field, which is the result of the conservation of spatial symmetries.

AB-stacked trilayer graphene possesses a slightly separated Dirac cone structure and two pairs of parabolic bands, including the superposition of monolayer- and bilayer-like energy bands. The optical excitations vanish between the former and the latter. Only an observable shoulder structure due to the distinct parabolic bands appears in the low-frequency absorption spectrum, while there exist five symmetric peaks at the middle frequency. Any excitation channel between the valence and conduction bands becomes available for a sufficiently high electric field. More low-frequency special structures are generated in the absorption spectra, in which the asymmetric peaks are related to the E_0-dependent oscillatory and sombrero-shaped bands with circular constant energy contours. However, the optical threshold frequency remains at zero, i.e. the semimetal–semiconductor transition is absent during the variation of the electric field. Similar behavior can also be observed in the AB stacking of $N > 3$. In general, the LL splitting energies are negligible, so the magneto-absorption spectra are almost the same for the distinct localization centers. One category along with four other types of magneto-optical excitation, which, respectively, come from monolayer- and bilayer-like LLs, exhibit many single- and twin-peak structures. Such absorption peaks due to the well-behaved modes obey the specific selection rules. The former is much stronger than the latter; furthermore, a simple relationship between the intensity and absorption frequency is absent for the latter. In particular, the certain initial excitations associated with the double-degenerate LLs of ($n_1^{c,v} = 0$; $n_2^{c,v} = 0$ and 1) can survive and are distinguishable in the single-peak absorption frequencies. The magneto-optical spectra are greatly enriched by the electric field because of the splitting LLs and the intragroup LL anticrossings. Five categories of inter-LL excitations change into nine. The twin peaks and the double peaks can occur simultaneously, reflecting the asymmetric and the splitting LL energy spectra, respectively. Moreover, the frequent anticrossings of the perturbed LLs in the E_0-dependent energy spectrum lead to an increase in peaks and reduced intensities.

ABC-stacked trilayer graphene has three pairs of unusual energy bands, namely partially flat (surface-localized), sombrero-shaped and linear. Excitation channels between any valence and conduction band are available. The threshold frequency is zero and the first absorption peak corresponds to the band-edge state excitations between the surface-localized and sombrero-shaped energy bands. There are four obvious peaks in the middle-frequency absorption spectrum. The optical gap (energy gap), which is characterized by the threshold peak from the separated surface-localized bands, is created by an electric field. The original peak exhibits an obvious blue shift, and three groups of LLs can generate nine categories of absorption peaks, including six intergroup and three intragroup types. Remarkably, intergroup LL anticrossings frequently emerge in the B_0-dependent energy spectrum and thus induce extra absorption peaks. Many of the prominent absorption peaks associated with the well-behaved LLs are presented under specific selection rules. As for these two kinds of absorption peaks, their B_0 dependence is rather strong. The magneto-optical properties are diversified by the electric field. With an increased electric field strength, many outstanding inter-LL absorption peaks gradually change from twin-peak into double-peak structures, as a result of the non-equivalence condition imposed on the two sublattices with identical projections. In particular, a single threshold peak becomes a double-peak structure, owing to the Fermi–Dirac distribution. The extra peaks are generated by the intragroup LL anticrossings, and the abnormal E_0-dependent absorption frequencies are simultaneously revealed.

The AAB-stacked trilayer graphene, without the z-plane mirror symmetry, sharply contrasts with the other highly symmetric stacking systems. It is a narrow-gap semiconductor, while the others are semimetals. The electronic and optical properties are enriched by the various interlayer atomic interactions. The unique band structure includes oscillatory, sombrero-shaped and parabolic energy dispersions. The 3×3 excitation channels present more special structures in the low- and middle-frequency optical spectrum, compared to those of the ABC stacking. As for the first asymmetric absorption peak, its frequency has an energy gap between the valence and conduction oscillatory bands; furthermore, both the frequency and intensity are significantly altered by the electric field. The LLs are split by the broken mirror symmetry, which is a phenomenon that is absent in other stacking systems. Intragroup and intergroup anticrossings occur frequently because of the complex relationship between the magnetic quantization and the interlayer atomic interactions. There are nine categories of absorption peaks, in which the intergroup LL excitations are much higher than those of the intragroup, depending on the phases and amplitudes of the LL wavefunctions. The magneto-absorption spectra exhibit single- or twin-peak structures which are dominated by the quantum modes, LL energy spectra and Fermi–Dirac distribution. The splitting LLs, with different localization centers, can generate very distinct absorption spectra, including extra single peaks during the anticrossings. A lot of strong peaks, which arise from well-behaved LLs, can be characterized by the specific selection rules. The intensity and frequency of the magneto-absorption peaks are strongly affected by the electric field.

Various stacking configurations can diversify the electronic structures and absorption spectra in sliding BLG. The vertical or non-vertical Dirac cones transform into parabolic bands, accompanied by the separation of cone structures, the creation of an arc-shaped stateless region, distorted energy dispersions, extra low-energy critical points and the splitting of middle-energy states. All the stacking systems belong to gapless semimetals, while an optical gap is only present in the AA and AA′ stacking because of the intracone excitations. The low-frequency shoulders or peaks strongly depend on the stacking configuration; furthermore, there are three or four middle-frequency strong peaks in the absence of three-fold rotation symmetry. An electric field causes the semimetallic BLG to become a semiconductor, except for AA and AA′ types; therefore, optical gaps are revealed in any absorption spectrum. This also leads to extra-low-frequency peaks related to the seriously distorted or weakly dispersive energy bands. As for the magneto-optical spectra, the diverse selection rules are dominated by three kinds of LLs with normal or irregular spatial distributions. A lot of strong absorption peaks due to the well-behaved LLs obey specific selection rules, such as ($\Delta n = \pm 1$; ± 1; ± 1, 0, ± 2) for AA, AB and AA′ stacking. Two novel selection rules, $\Delta n = 0$ and $\pm$ 2, are verified to arise from the tilted Dirac cones in the AA′ stacking. The anticrossing perturbated LLs induce extra selection rules ($\Delta n = 0$ and ± 2) in the intergroup optical excitations, but the heights of the absorption peaks are relatively low. More peculiarly, the undefined LLs create a large number of absorption peaks in the absence of a specific selection rule. Magneto-optical spectra in a composite field are more complicated as they involve drastic changes in the number and intensity of absorption peaks.

Up to now, experimental observations have been consistent with part of the theoretical calculations on the optical properties in external fields. As for AB-stacked bilayer graphene, the ~0.3 eV shoulder structure under a zero field, the **E**-induced semimetal–semiconductor transition and two low-frequency asymmetric peaks, the two strong π-electronic absorption peaks at the middle frequency, the $\Delta n = \pm 1$ selection rule for the first group of LLs and the linear B_0 dependence of the inter-LL excitation energies have been confirmed by absorption, transmission, reflection, and/or Raman scattering spectroscopy. Similar verification performed on trilayer ABA stacking contain one shoulder at ~0.5 eV, gapless behavior unaffected by the electric field, **E**-generated low-frequency multipeak structures, several π-electronic peaks, and monolayer- and bilayer-like inter-LL absorption frequencies. Moreover, the identified spectral features in ABC-stacked trilayer graphene cover two low-frequency characteristic peaks and a gap opening with the electric field. However, plenty of important predicted results require further experimental examination, including the shoulders, asymmetric and symmetric peaks in zero-field optical spectra, the initial forbidden excitation region in even-layer AA stacking systems, the more spectral structures in lower-symmetry stacking configurations, **E**-created optical gaps and special structures, the significant dependence of the prominent π-electronic peaks on the stacking configuration and layer number, the diverse magneto-optical selection rules due to three kinds of LLs, the delta-function-like inter-LL absorption peaks with single, twin and double forms, the almost uniform single peaks related to the non-tilted Dirac cones, the LL-anticrossing-induced drastic changes in the number and intensity of the peaks and the greatly diversified magneto-absorption spectra under a composite field. The experimental measurements on the frequency and form of spectral structures can provide critical information for identifying energy bands and LL energy spectra, such as the linear, parabolic, sombrero-shaped and oscillatory dispersions and the symmetric, asymmetric and splitting LL energy spectra. They are very useful in understanding the complex relationship between the stacking configurations, the magnetic quantization and the Coulomb potential energy. They could also be used to distinguish the geometric structures of the distinct stacking domains.

In addition to the stacking configurations, the different dimensionalities can create diverse properties in carbon-related systems. One-dimensional graphene nanoribbons [14-29], 2D graphene and 3D graphite [30–44] all differ quite significantly from one another in their electronic and optical properties. Concerning the first system, the finite-size quantum confinement and edge structure dominate the essential properties. A lot of asymmetric absorption peaks, which come from the 1D parabolic energy bands in the absence of external fields, obey the armchair- or zigzag-dependent selection rule [23–26]. The spatial distributions of the standing waves are responsible for this unusual rule, since they are mainly determined by distinct boundary structures. When the ribbon width is longer than the magnetic length, each 1D Landau subband consists of dispersionless quasi-LL and parabolic dispersion. The almost well-behaved quasi-LLs are localized at the ribbon center and they exhibit prominent symmetric peaks with a selection rule similar to that in monolayer graphene [27]. The edge- and magneto-dependent selection rules will be destroyed by a sufficiently high transverse electric field [28, 29]. On the other hand, the periodic interlayer atomic interactions play a critical role in 3D graphite; AA- [41], AB- [30–32, 42] and ABC-stacked [43] graphite, respectively, have one pair, two pairs and one pair of 3D valence and conduction bands. Their low-frequency absorption spectra show a shoulder structure after the interband forbidden region, a concave-downward structure and a shoulder structure, respectively. Specifically, the optical gaps vanish in the latter two systems [33, 34]. The 3D magnetically quantized states form Landau subbands (LSs) composed of 2D dispersionless LLs and 1D k_z-dependent energy dispersions. The number of LS groups is, respectively, one, two and one for simple hexagonal, Bernal and rhombohedral graphite. Many LSs in AA-stacked graphite cross the Fermi level, and their bandwidths are $\sim$1 eV, reflecting the strongest interlayer atomic interactions between the three systems. Such LSs generate plenty of twin asymmetric peak structures in opposite directions [35, 36]. The LSs of AB-stacked graphite possess $\sim$0.2 eV band widths, and only two of them intersect with E_F [37, 38, 44]. The LS energy spectrum corresponding to the K and H points (k_z = 0 and 1) is, respectively, almost asymmetric and symmetric about E_F, so that the magneto-absorption spectrum presents the twin structures of asymmetric peaks in the same direction and single symmetric peaks [38, 42]. As for the ABC-stacked graphite, the LS bandwidths are narrower than 10 meV, and only one is located at E_F [39]. These result in a lot of single absorption peaks with symmetric forms, which are attributed to the merged excitations from the K and H points [40]. It is also noticed that the LS energies of AA- and ABC-stacked graphite [35, 39] are characterized by a similar square root dependence on the quantum number and the magnetic field strength. A review paper on the electronic and optical properties has been finished for 1D graphene nanoribbons [29], and it will be extended to 3D graphite.

2D materials have attracted much attention since the discovery of graphene in 2004. Graphene-related materials have shown great potential for photopic and optoelectronic applications, such as touch-screen panel devices [45], light-emitting diodes [46], solar cells [47], photo-detectors [48, 49] and photo-modulators [50, 51]. The high transparency and flexibility of graphene can be used to design complicated and thin devices. Compared with semiconductor optical modulators, graphene-based modulators have the appealing advantage of having a low operation voltage, fast modulation speed, small footprint and large optical bandwidth. By tuning the

Fermi level or the free carrier concentration of the graphene sheet, the modulator is operated over a broad wavelength range. Furthermore, because of their much stronger interband optical transition, detectors based on graphene enable applications in the energy spectrum over a very wide range, including ultraviolet, visible, infrared and terahertz frequencies.

Layered group IV condensed matter systems with nanoscaled thickness and hexagonal symmetry are very suitable for studying novel physical, chemical and material phenomena [52–58]. Hybrid systems based on a combination of different 2D systems, e.g. graphene and other group IV materials, have great potential for future technological applications [59–61]. Silicene and germanene have been synthesized on metallic substrates [52–57], and recently, tinene was successfully fabricated on a substrate of bismuth telluride [58]. Silicene, germanene and tinene possess a buckled structure with mixed $sp^2 - sp^3$ orbital hybridization rather than just sp^2. Their spin–orbital interactions are sufficiently strong for them to be able to create an observable energy gap. The sp^3 orbital bonding and the spin–orbital interactions can dramatically change the electronic structures; therefore, they are expected to create rich magnetic quantization.

The generalized tight-binding model could be further modified to comprehend diverse quantization phenomena in various condensed matter systems. Recently, a new Hamiltonian was proposed to explore the unusual magnetic properties of monolayer tinene [62]. The magnetic quantization of the multi-orbital-, spin- and configuration-dependent Hamiltonian was solved by using the generalized tight-binding model. This will be very helpful in the further development of critical Hamiltonian models in mainstream materials. The Hamiltonian dimension of monolayer tinene is eight times as large as that of monolayer graphene. The buckled structure causes the matrix elements to be complex at any wave vector, so that Hamiltonian diagonalization becomes difficult. Tinene is predicted to have two groups of low-lying LLs dominated by the $5p_z$ and ($5p_x$, $5p_y$) orbitals, indicating the critical role of multiorbital bonding and spin–orbital coupling. The important differences between tinene and graphene in the main characteristics of LLs include the orbital components, spin configurations, localization centers, state degeneracy, and magnetic- and electric-field dependencies. Other quantized properties are worthy of a detailed investigation, such as the optical spectra, Coulomb excitations and transport properties of layered group IV systems.

References

[1] Foa Torres L E F, Perez-Piskunow P M, Balseiro C A and Usaj G 2014 Multiterminal conductance of a Floquet topological insulator Phys. Rev. Lett. **113** 266801

[2] Ezawa M 2012 A topological insulator and helical zero mode in silicene under an inhomogeneous electric field *New J. Phys.* **14** 033003

[3] Ezawa M 2012 Valley-polarized metals and quantum anomalous Hall effect in silicene *Phys. Rev. Lett.* **109** 055502

[4] Tai Y Y, Choi H C, Ahmed T, Ting C S and Zhu J X 2015 Edge states and local electronic structure around an adsorbed impurity in a topological superconductor *Phys. Rev.* B **92** 174514

[5] Li M Y, Shi Y, Cheng C C, Lu L S, Lin Y C and Tang H L *et al* 2015 Epitaxial growth of a monolayer WSe_2-MoS_2 lateral p-n junction with an atomically sharp interface *Science* **349** 524–58
[6] Ho Y H, Wang Y H and Chen H Y 2014 Magnetoelectronic and optical properties of a MoS_2 monolayer *Phys. Rev.* B **89** 155316
[7] Ho Y H, Chiu C W and Lin M F 2014 Magneto-optical spectra of transition metal dichalcogenides: a comparative study *Appl. Phys. Lett.* **105** 222411
[8] Tabert C J and Nicol E J 2013 Valley-spin polarization in the magneto-optical response of silicene and other similar 2D crystals *Phys. Rev. Lett.* **110** 197402
[9] Tabert C J and Nicol E J 2013 Magneto-optical conductivity of silicene and other buckled honeycomb lattices *Phys. Rev.* B **88** 085434
[10] Schafgans A A, Post K W, Taskin A A, Ando Y, Qi X L and Chapler B C *et al* 2012 Landau level spectroscopy of surface states in the topological insulator $Bi_{0.91}Sb_{0.09}$ via magneto-optics *Phys. Rev.* B **85** 195440
[11] LaForge A D, Frenzel A, Pursley B C, Lin T, Liu X F and Shi J *et al* 2010 Optical characterization of Bi_2Se_3 in a magnetic field: infrared evidence for magnetoelectric coupling in a topological insulator material *Phys. Rev.* B **81** 125120
[12] Morkoc H, Strite S, Gao G B, Lin M E, Sverdlov B and Burns M 1994 Large-band-gap SiC, III-V nitride, and II-VI ZnSe-based semiconductor-device technologies *J. Appl. Phys* **76** 1363–98
[13] Ando T, Fowler A B and Stern F 1982 Electronic properties of two-dimensional systems *Rev. Mod. Phys.* **54** 437–72
[14] Li X L, Wang X R, Zhang L, Lee S W and Dai H J 2008 Chemically derived, ultrasmooth graphene nanoribbon semiconductors *Science* **319** 1229–32
[15] Fujii S and Enoki T 2010 Cutting of oxidized graphene into nanosized pieces *J. Am. Chem. Soc.* **132** 10034–41
[16] Kosynkin D V, Higginbotham A L, Sinitskii A, Lomeda J R, Dimiev A and Price B K *et al* 2009 Longitudinal unzipping of carbon nanotubes to form graphene nanoribbons *Nature* **458** 872–U5
[17] Jiao L, Zhang L, Wang X, Diankov G and Dai H J 2009 Narrow graphene nanoribbons from carbon nanotubes *Nature* **458** 877–80
[18] Kim K, Sussman A and Zettl A 2010 Graphene nanoribbons obtained by electrically unwrapping carbon nanotubes ACS Nano **4** 1362–6
[19] Blankenburg S, Cai J, Ruffieux P, Jaafar R, Passerone D and Feng X *et al* 2012 Intraribbon heterojunction formation in ultranarrow graphene nanoribbons *ACS Nano* **6** 2020–5
[20] Yi Z, Yanfang Z, Geng L, Jianchen L, Xiao L and Shixuan D *et al* 2014 Direct visualization of atomically precise nitrogen-doped graphene nanoribbons *Appl. Phys. Lett.* **105** 023101
[21] Xie L M, Wang H L, Jin C H, Wang X R, Jiao L Y and Suenaga K *et al* 2011 Graphene nanoribbons from unzipped carbon nanotubes: atomic structures, Raman spectroscopy, and electrical properties *J. Am. Chem. Soc.* **133** 10394–7
[22] Chang S L, Wu B R, Wong J H and Lin M F 2014 Configuration-dependent geometric and electronic properties of bilayer graphene nanoribbons *Carbon* **77** 1031–9
[23] Lin M F and Shyu F L 2000 Optical properties of nanographite ribbons *J. Phys. Soc. Jpn* **69** 3529–32
[24] Hsu H and Reichl L E 2007 Selection rule for the optical absorption of graphene nanoribbons *Phys. Rev.* B **76** 045418

[25] Chung H C, Lee M H, Chang C P and Lin M F 2011 Exploration of edge-dependent optical selection rules for graphene nanoribbons *Opt. Express* **19** 23350–63

[26] Sasaki K I, Kato K, Tokura Y, Oguri K and Sogawa T 2010 Theory of optical transitions in graphene nanoribbons *Phys. Rev.* B **84** 085458

[27] Huang Y C, Chang C P and Lin M F 2007 Magnetic and quantum confinement effects on electronic and optical properties of graphene ribbons *Nanotechnology* **18** 495401

[28] Chang C P, Huang Y C, Lu C L, Ho J H, Li T S and Lin M F 2006 Electronic and optical properties of a nanographite ribbon in an electric field *Carbon* **44** 508–15

[29] Chung H C, Chang C P, Lin C Y and Lin M F 2016 Electronic and optical properties of graphene nanoribbons in external fields *Phys. Chem. Chem. Phys.* **18** 7573–616

[30] Soule D E 1958 Magnetic field dependence of the Hall effect and magnetoresistance in graphite single crystals *Phys. Rev.* **112** 698–707

[31] Galt J K, Yager W A and Dail H W 1956 Cyclotron resonance effects in graphite *Phys. Rev.* **103** 1586–7

[32] Dresselhaus M S and Dresselhaus G 1981 Intercalation compounds of graphite *Adv. Phys.* **30** 139–326

[33] Chiu C W, Lee S H, Chen S C, Shyu F L and Lin M F 2010 Absorption spectra of AA-stacked graphite *New J. Phys.* **12** 083060

[34] Chiu C W, Huang Y C, Chen S C, Lin M F and Shyu F L 2011 Low-frequency electronic and optical properties of rhombohedral graphite *Phys. Chem. Chem. Phys.* **13** 6036–42

[35] Chen R B, Chiu Y H and Lin M F 2013 A theoretical evaluation of the magneto-optical properties of AA-stacked graphite *Carbon* **54** 268–76

[36] Chen R B, Chiu Y H and Lin M F 2015 Beating oscillations of magneto-optical spectra in simple hexagonal graphite *Comput. Phys. Commun.* **189** 60–5

[37] Nakao K 1976 Landau level structure and magnetic breakthrough in graphite *J. Phys. Soc. Jpn* **40** 761–8

[38] Ho Y H, Wang J, Chiu Y H, Lin M F and Su W P 2011 Characterization of Landau subbands in graphite: a tight-binding study *Phys. Rev.* B **83** 121201

[39] Ho C H, Chang C P and Lin M F 2013 Precessing anisotropic Dirac cone and Landau subbands along a nodal spiral *New J. Phys.* **15** 053032

[40] Ho C H, Chang C P and Lin M F 2015 Optical magnetoplasmons in rhombohedral graphite with a three-dimensional Dirac cone structure *J. Phys. Condens. Matter* **27** 125602

[41] Lee J K, Lee S C, Ahn J P, Kim S C, Wilson J I B and John P 2008 The growth of AA graphite on (1 1 1) diamond *J. Chem. Phys.* **129** 234709

[42] Orlita M, Faugeras C, Schneider J M, Martinez G, Maude D K and Potemski M 2009 Graphite from the viewpoint of Landau level spectroscopy: An effective graphene bilayer and monolayer *Phys. Rev. Lett.* **102** 166401

[43] Lipson H and Stokes A R 1942 The structure of graphite *Proc. R. Sot. London, Ser.* A **181** 101

[44] Koshino M and Ando T 2008 Magneto-optical properties of multilayer graphene *Phys. Rev.* B **77** 115313

[45] Bae S, Kim H, Lee Y, Xu X, Park J S and Zheng Y *et al* 2010 Roll-to-roll production of 30-inch graphene films for transparent electrodes *Nat. Nanotechnol.* **5** 574–8

[46] Matyba P, Yamaguchi H, Eda G, Chhowalla M, Edman L and Robinson N D 2010 Graphene and mobile ions: the key to all-plastic, solution-processed light-emitting devices *ACS Nano* **4** 637–42

[47] Kusumawati Y, Koussi-Daoud S and Pauporte T 2016 TiO_2/graphene nanocomposite layers for improving the performances of dye-sensitized solar cells using a cobalt redox shuttle *J. Photochem. Photobiol.* **329** 54–60
[48] Xia F N, Mueller T, Lin Y M, Valdes-Garcia A and Avouris P 2009 Ultrafast graphene photodetector *Nat. Nanotechnol.* **4** 839–43
[49] Mueller T, Xia F and Avouris P 2010 Graphene photodetectors for high-speed optical communications *Nat. Photon.* **4** 297–301
[50] Liu M, Yin X B, Ulin-Avila E, Geng B S, Zentgraf T and Ju L *et al* 2011 A graphene-based broadband optical modulator *Nature* **474** 64–7
[51] Liu J B, Li P J, Chen Y F, Song X B, Mao Q and Wu Y *et al* 2016 Flexible terahertz modulator based on coplanar-gate graphene field-effect transistor structure *Optics Lett* **41** 816–9
[52] Aufray B, Kara A, Vizzini S, Oughaddou H, Leandri C and Ealet B *et al* 2010 Graphene-like silicon nanoribbons on Ag(110): a possible formation of silicene *Appl. Phys. Lett.* **96** 183102
[53] Lalmi B, Oughaddou H, Enriquez H, Kara A, Vizzini S and Ealet B *et al* 2010 Epitaxial growth of a silicene sheet *Appl. Phys. Lett.* **97** 223109
[54] Vogt P, de Padova P, Quaresima C, Avila J, Frantzeskakis E and Asensio M C *et al* 2012 Silicene: compelling experimental evidence for graphenelike two-dimensional silicon *Phys. Rev. Lett.* **108** 155501
[55] Bianco E, Butler S, Jiang S, Restrepo O, Windl W and Goldberger J 2013 Stability and exfoliation of germanane: a germanium graphane analogue *ACS Nano* **7** 4414–21
[56] Chu J H, Kwak J, Kim S D, Lee M J, Kim J J and Park S D *et al* 2014 Monolithic graphene oxide sheets with controllable composition *Nature Comm.* **5** 3383
[57] Derivaz M, Dentel D, Stephan R, Hanf M C, Mehdaoui A and Sonnet P *et al* 2015 Continuous germanene layer on A(111) *Nano Lett.* **15** 2510–6
[58] Zhu F F, Chen W J, Xu Y, Gao C L, Guan D D and Liu C H *et al* 2015 Epitaxial growth of two-dimensional stanene *Nat. Mater.* **14** 1020
[59] Sun Z, Hasan T, Torrisi F, Popa D, Privitera G and Wang F *et al* 2010 Graphene mode-locked ultrafast laser *ACS Nano* **4** 803–10
[60] Tao L, Cinquanta E, Chiappe D, Grazianetti C, Fanciulli M and Dubey M *et al* 2015 Silicene field-effect transistors operating at room temperature *Nat. Nanotech.* **10** 227–31
[61] Tsai W F, Huang C Y, Chang T R, Lin H, Jeng H T and Bansil A 2013 Gated silicene as a tunable source of nearly 100 percent spin-polarized electrons *Nat. Commun.* **4** 1500
[62] Chen S C, Wu C L, Wu J Y and Lin M F 2016 Novel magnetic quantization of sp^3 bonding in monolayer tinene *Phys. Rev.* B **94** 045410

Lightning Source UK Ltd.
Milton Keynes UK
UKHW05n2010260618
324811UK00003B/58/P

9 780750 315647